MATHEMATICS
SUCCESS MANTRA

The complete formulae book useful for JEE Mains, Advanced, BITS, MHT CET, Karnataka CET, AP EAPCET, Telangana EAMCET, RRB, SSC and all other competitive exams

By

Ravi Kumar Kothapalli

ALL RIGHTS RESERVED

All rights reserved. No part of this publication may be reproduced, stored in or introduced into a retrieval system, or transmitted, in any form by any means may it be electronically, mechanical, optical, chemical, manual, photocopying, or recording without prior written permission of the Publisher/ Author.

MATHEMATICS
SUCCESS MANTRA

by

Ravi Kumar Kothapalli

Copy Right: Ravi Kumar Kothapalli

Published By: Kasturi Vijayam

Published on: Jul/2024

ISBN (Paperback): 978-81-974474-4-0

Print On Demand

Ph:0091-9515054998

Email: Kasturivijayam@gmail.com

Book Available

@

Amazon, flipkart

Author's Note

"I aspire to be a good human teacher who possesses wisdom to teach my students, and show them that I genuinely care, and am concerned with their learning, a good human teacher who cannot be replaced by technology. As the old saying goes... "Give a man a fish, and you feed him for a day."

I identify the despite of efforts; some students struggle to secure good results. I believe that the key to success lies in applying hard work correctly. The plethora of books which are available today can often confuse students, they are putting them in a tough spot. Are these books truly beneficial? It's doubtful.

After much consideration, I've decided to publish a book that caters to all types of students. I have intentionally focused on presenting fundamental ideas and equations. Mastering these basic concepts and principles will help students achieve top grades. However, in this competitive environment, securing good grades is crucial.

Here are my recommendations for problem-solving:
- * Read the problem thoroughly;
- * Understand what is being asked of you;
- * Gather relevant concepts and formulae;
- * Apply them to the problem;
- * Draw diagrams when necessary.

I believe that fundamental ideas, rules, and equations are essential for solving most problems. This is why I introduced the book 'Mathematics Success Mantra.' I hope 'The Success Mantra' will be well-received and beneficial to all students.

I dedicate this book to Sri J.R.L.V. Sharma, my math teacher, who has been a significant influence on my life. He transformed me from an average student into a successful math teacher, and Ravi Kumar kothapalli, the author of maths books.

I have worked diligently to ensure this book is error-free. However, some printing errors may still be present. I would greatly appreciate it if you could inform me of any errors you find. We welcome constructive feedback and suggestions for improvement, which will be included in the next edition.

- **Ravi Kumar Kothapalli**

Contents

ALGEBRA 1

    1. BASIC FORMULAE OF ALGEBRA   2

    2. LAWS OF INDICES 3

    3. SURDS 4

    4. LOGARITHMS 5

    5. DIVISIBILITY RULES 6

    6. SETS 8

    7. RELATIONS 11

    8. FUNCTIONS 13

    9. MATHEMATICAL REASONING 20

    10. SEQUENCES AND SERIES 22

    11. MATRICES 26

    12. DETERMINANTS 31

    13. QUADRATIC EQUATIONS AND EXPRESSIONS 35

    14. THEORY OF EQUATIONS 40

    15. PERMUTATIONS AND COMBINATIONS 44

    16. BINOMIAL THEOREM 53

TRIGONOMETRY 57

    17. TRIGONOMETRIC RATIOS AND IDENTITIES 58

    18. PERIODICITY AND EXTREME VALUES 63

    19. COMPOUND ANGLES 65

    20. MULTIPLE AND SUB-MULTIPLE ANGLES 67

    21. TRANSFORMATIONS 70

    22. TRIGONOMETRIC EQUATIONS 71

    23. INVERSE TRIGONOMETRIC FUNCTIONS 72

    24. PROPERTIES OF TRIANGLES 75

    25. HEIGHTS AND DISTANCES 79

    26. HYPERBOLIC FUNCTIONS 83

    27. COMPLEX NUMBERS AND DEMOIVRE'S THEOREM 86

2-DIMENSIONAL GEOMETRY 93

    28. 2D CO-ORDINATES 94

    29. LOCUS 100

    30. CHANGE OF AXES 101

    31. STRAIGHT LINES 103

    32. PAIR OF STRAIGHT LINES 110

33. CIRCLES 114

34. SYSTEM OF CIRCLES 124

35. CONIC SECTION 126

36. PARABOLA 128

37. ELLIPSE 134

38. HYPERBOLA 139

3D GEOMETRY **145**

39. 3D CO-ORDINATES 146

40. DIRECTION COSINES AND DIRECTION RATIOS 149

41. 3D PLANES 151

42. 3D LINES 154

VECTOR ALGEBRA **156**

43. ADDITION OF VECTORS 157

44. MULTIPLICATION OF VECTORS 163

CALCULUS **170**

45. LIMITS, CONTINUITY AND DIFFERENTIABILITY 171

46. DIFFERENTIATION 177

47. APPLICATIONS OF DERIVATIVES 179

48. INDEFINITE INTEGRATION 186

49. DEFINITE INTEGRALS 191

50. AREAS 193

51. DIFFERENTIAL EQUATIONS 195

STATISTICS **198**

52. MEASURES OF DISPERSION 199

53. PROBABILITY 204

54. RANDOM VARIABLES AND PROBABILITY DISTRUBUTIONS 208

ALGEBRA

1. BASIC FORMULAE OF ALGEBRA

1. $(a+b)^2 = a^2 + 2ab + b^2$
2. $(a-b)^2 = a^2 - 2ab + b^2$
3. $(a+b)^2 + (a-b)^2 = 2(a^2 + b^2)$
4. $(a+b)^2 - (a-b)^2 = 4ab$
5. $a^2 + b^2 = (a+b)^2 - 2ab = (a-b)^2 + 2ab$
6. $(a+b)^2 = (a-b)^2 + 4ab$
7. $(a-b)^2 = (a+b)^2 - 4ab$
8. $a^2 - b^2 = (a+b)(a-b)$
9. $(a+b)^3 = a^3 + 3a^2b + 3ab^2 + b^3 = a^3 + b^3 + 3ab(a+b)$
10. $(a-b)^3 = a^3 - 3a^2b + 3ab^2 - b^3 = a^3 - b^3 - 3ab(a-b)$
11. $a^3 + b^3 = (a+b)^3 - 3ab(a+b) = (a+b)(a^2 - ab + b^2)$
12. $a^3 - b^3 = (a-b)^3 + 3ab(a-b) = (a-b)(a^2 + ab + b^2)$
13. $(a+b+c)^2 = a^2 + b^2 + c^2 + 2ab + 2bc + 2ca$
    $= a^2 + b^2 + c^2 + 2(ab + bc + ca)$
14. $a^2 + b^2 + c^2 = (a+b+c)^2 - 2(ab + bc + ca)$
15. If $a^2 + b^2 + c^2 = 0$ then $a = b = c = 0$
16. $a^3 + b^3 + c^3 = (a+b+c)(a^2 + b^2 + c^2 - ab - bc - ca)$
    $$= \frac{1}{2}(a+b+c)[(a-b)^2 + (b-c)^2 + (c-a)^2]$$
17. If $a+b+c = 0$ then $a^3 + b^3 + c^3 = 3abc$
18. If $a^3 + b^3 + c^3 = 3abc$ then $a+b+c = 0$ or $a = b = c = 0$
19. If $a + \frac{1}{a} = k$ then $a^2 + \frac{1}{a^2} = k^2 - 2$
20. If $a - \frac{1}{a} = k$ then $a^2 + \frac{1}{a^2} = k^2 + 2$
21. If $a + \frac{1}{a} = k$ then $a^3 + \frac{1}{a^3} = k^3 - 3k$
22. If $a - \frac{1}{a} = k$ then $a^3 - \frac{1}{a^3} = k^3 + 3k$
23. $x^4 + x^2 + 1 = (x^2 + x + 1)(x^2 - x + 1)$
24. $ab + bc + ca = abc\left(\frac{1}{a} + \frac{1}{b} + \frac{1}{c}\right)$
25. If $ab + bc + ca = 0$ then $abc = 0$ or $\frac{1}{a} + \frac{1}{b} + \frac{1}{c} = 0$
26. **Componendo and dividendo rule:**

    If $\frac{a}{b} = \frac{c}{d}$ then $\frac{a+b}{a-b} = \frac{c+d}{c-d}$ or $\frac{a-b}{a+b} = \frac{c-d}{c+d}$ or $\frac{b+a}{b-a} = \frac{d+c}{d-c}$ or $\frac{b-a}{b+a} = \frac{d-c}{d+c}$
27. If $a > b > c$ then $-a < -b < -c$
28. If $a > b > c$ then $\frac{1}{a} < \frac{1}{b} < \frac{1}{c}$
29. If $|x| < a$ then $-a < x < a$
30. If $|x| > a$ then $x < -a$ or $x > a$

2. LAWS OF INDICES

1. $a \times a \times a \times \ldots\ldots\ldots n \text{ times} = a^n$
2. $a^m \times a^n = a^{m+n}$
3. $\dfrac{a^m}{a^n} = a^{m-n}$
4. $a^{-m} = \dfrac{1}{a^m}$
5. $(a^m)^n = (a^n)^m = a^{mn}$
6. $(ab)^n = a^n b^n$
7. $\left(\dfrac{a}{b}\right)^n = \dfrac{a^n}{b^n}$
8. $\left(\dfrac{a}{b}\right)^{-n} = \left(\dfrac{b}{a}\right)^n$
9. If $a^m = a^n$ then $m = n$
10. If $a^m = b^m$ then $a = b$
11. $\sqrt[n]{a} = a^{1\backslash n}$
12. $\sqrt[n]{ab} = (ab)^{1\backslash n}$
13. $\sqrt[n]{\dfrac{a}{b}} = \left(\dfrac{a}{b}\right)^{1\backslash n} = \left(\dfrac{b}{a}\right)^{-1\backslash n}$
14. $\left(\sqrt[n]{a}\right)^m = (a)^{m\backslash n}$

3. SURDS

1. Conjugate of $a + \sqrt{b}$ is $a - \sqrt{b}$
2. Conjugate of $a - \sqrt{b}$ is $a + \sqrt{b}$
3. $\sqrt{a + b + 2\sqrt{ab}} = \sqrt{a} + \sqrt{b}$
4. $\sqrt{a + b - 2\sqrt{ab}} = \begin{cases} \sqrt{a} - \sqrt{b} \text{ when } a > b \\ \sqrt{b} - \sqrt{a} \text{ when } b > a \end{cases}$
5. $\sqrt{a\sqrt{a\sqrt{a\sqrt{a \ldots \ldots \infty}}}} = a$
6. $\sqrt{a\sqrt{a\sqrt{a\sqrt{a \ldots \ldots n \text{ times}}}}} = a^{1-\frac{1}{2^n}}$
7. $\sqrt[n]{a\sqrt[n]{a\sqrt[n]{a\sqrt[n]{a \ldots \ldots \infty}}}} = a^{\frac{1}{n-1}}$
8. $\sqrt{a + \sqrt{a + \sqrt{a + \sqrt{a \ldots \ldots \infty}}}} = \frac{\sqrt{4a+1}+1}{2}$
9. $\sqrt{a - \sqrt{a - \sqrt{a - \sqrt{a \ldots \ldots \infty}}}} = \frac{\sqrt{4a+1}-1}{2}$
10. $\sqrt{a + \sqrt{a - \sqrt{a + \sqrt{a - \cdots \ldots \infty}}}} = \frac{\sqrt{4a-3}+1}{2}$
11. $\sqrt{a - \sqrt{a + \sqrt{a - \sqrt{a + \cdots \ldots \infty}}}} = \frac{\sqrt{4a-3}-1}{2}$

4. LOGARITHMS

1. $\log_a 1 = 0$
2. $\log_a a = 1$
3. $\log_b a = \frac{\log_e a}{\log_e b}$
4. $\log_b a = \log_c a \cdot \log_b c$
5. $\log_b a = \frac{1}{\log_a b}$
6. $\log_b a^n = n \log_b a$
7. $\log_{b^n} a = \frac{1}{n} \log_b a$
8. $a^{\log_a x} = x$
9. $a^{\log_c b} = b^{\log_c a}$
10. $\log_e (mn) = \log_e m + \log_e n$
11. $\log_e \left(\frac{m}{n}\right) = \log_e m - \log_e n$
12. If $a^b = x$ then $b = \log_a x$
13. If $\log_b a > x$ then $a > b^x$ when $b > 1$
14. If $\log_b a > x$ then $a < b^x$ when $0 < b < 1$

5. DIVISIBILITY RULES

1. If the last digit of a number is an even number then the number is divisible by 2.
   Eg: 243576
   The last digit of the number is 6 which is an even number. So, the number is divisible by 2.

2. If the sum of the all the digits of a number is divisible by 3, then the number is also divisible by 3.
   Eg: 54303
   The sum of the digits= $5 + 4 + 3 + 0 + 3 = 15$ is divisible by 3, so the number is divisible by 3.

3. If the last two digits of a number are divisible by 4, then the number is divisible by 4.
   Eg: 34564
   The last two digits of the number are 64 which are divisible by 4, so the number is divisible by 4.

4. If the last digit of a number is 0 or 5, then the number is divisible by 5.
   Eg: 13560 or 43765
   The last digit of the numbers 0 or 5, so the numbers are divisible by 5.

5. If a number is divisible by both 2 and 3, then the number is divisible by 6.
   Eg: 12306
   The number is divisible by both 2 and 3, so the number is divisible by 6.

6. If the difference between twice the units digit of a number and the remaining part of the number is divisible by 7, then the number is divisible by 7.
   Eg: 24017
   $2401 - 2(7) = 2401 - 14 = 2387 = 7 \times 341$ which is divisible by 7. So the number is divisible by 7.

7. If the last three digits of a number are divisible by 8, then the number is also divisible by 8.
   Eg: 24344
   The last three digits of the number 344 is divisible by 8, so the number is divisible by 8.

8. If the sum of the all digits of a number is divisible by 9, then the number is also divisible by 9.
   Eg: 36522

The sum of the digits= 3 + 6 + 5 + 2 + 2 = 18 which is divisible by 9. So the number is divisible by 9.

9. If the last digit of a number is 0, then the number is divisible by 10.
   Eg: 75420
   The last digit of the number is 0, so the number is divisible by 10.

10. If the difference of the sum the digits in odd places and the sum of the digits in even places of a number is divisible by 11, then the number is also divisible by 11.
    Eg: 37829
    The sum of the digits in odd places=20
    The sum of the digits in even places=9
    Ther difference=11 which is divisible by 11. So the number is divisible by 11

11. If a number is divisible by both 3 and 4, then the number is divisible by 12.
    Eg: 65724
    The number is divisible by both 3 and 4, so the number is divisible by 12.

12. If the last two digits of a number are divisible by 25, then the number is divisible by 25.
    Eg: 56475
    The last two digits of the number are 75 which are divisible by 25, so the number is divisible by 25.

6. SETS

Set: A collection of well-defined distinct objects is called a set.

The sets always denoted by capital letters like A, B, C... and the objects (elements) denoted by small letters like a,b,c,... They represented in the brackets {   }.

Roaster form (or) Tabular form: A set is described by listing all the elements separated by commas and enclosed within the brackets.

Eg: $A = \{0,2,4,6,8\}$

Set-builder form: A set is described by characterizing the property of its elements.

Eg: If $A = \{1,2,3,4,5\}$ then the set-builder form of A is $\left\{\dfrac{x}{x} is\ a\ natural\ number\ less\ than\ 6\right\}$

Cardinal number: The number of elements of a set is called as a cardinal number of the set.

Eg: If $A = \{0,1,2,3\}$ then the cardinal number of A is $n(A) = 4$

Types of sets

1. **Singleton set:** If a set has only one element then the set is called a singleton set.

   Eg: $\{a\}$

2. **Empty set:** If a set has no elements then the set is called an empty set.

   It is denoted by $\emptyset\ or\ \{\ \}$

3. **Finite set:** If a set contains the finite number of elements then the set is called as a finite set.

   Eg: $A = \{a, e, i, o, u\}$

4. **Infinite set:** If a set contains the infinite number of elements then the set is called as an infinite set.

   Eg: $A = \{1, 3, 5, 7, \ldots \ldots\}$

5. **Equal sets:** Two sets A and B are said to be equal when they have same elements.

6. **Equivalent Sets:** Two sets A and B are said to be equivalent when they have same number of elements.

   Eg: $A = \{1,3,5\}$ and $B = \{2,4,6\}$ are equivalent since $n(A) = n(B)$

7. **Universal set:** The set contains all sets in the given context is called as a universal set.

   It is denoted by U.

8. **Sub set:** A set A is said to be subset of a set B when each element of A is the element of B.

   It is denoted by $A \subseteq B$.

9. **Proper sub set:** A set A is a subset of B but $A \neq B$ then A is said to be proper subset of B.

   It is denoted by $A \subset B$.

10. **Super set :** If A is a subset of B then B is called as a superset of A.
11. **Power set:** The set of all the subsets of a set is called as power set.

    If $n(A) = n$ then the number of elements in power set is 2^n.

12. **Disjoint sets:** The sets A and B are said to be disjoint sets when $A \cap B = \emptyset$

Union of sets: The set of all the elements which are in either a set A or a set B is called as the union of the sets A and B. It is denoted by $A \cup B$.

i.e. $A \cup B = \left\{\frac{x}{x} \in A \text{ or } x \in B\right\}$

Intersection of sets: The set of all the elements which are in commonly both in the sets A and B is called as an intersection of the sets A and B. It is denoted by $A \cap B$.

i.e. $A \cap B = \left\{\frac{x}{x} \in A \text{ and } x \in B\right\}$

Difference of two sets: Let A and B be two non empty sets. Then A-B is called as

the difference of A and B and is denoted by $A - B = \left\{\frac{x}{x} \in A \text{ and } x \notin B\right\}$

Complement of a set: Let A be a non empty set. Then the complement of A is denoted as A^1 or A^c and is defined as A^1 or $A^c = \{x \backslash x \in U \text{ and } x \notin A\}$

i.e. A^1 or $A^c = U - A$.

Symmetric difference of sets: Let A and B be two non empty sets. Then the symmetric difference of A and B is denoted by $A \Delta B$ and is defined as

$A \Delta B = (A - B) \cup (B - A) \text{ or } (A \cup B) - (A \cap B)$

Cardinal properties: If A,B and C are finite sets and U be the finite universal set then

1. $n(A - B) = n(A) - n(A \cap B)$
2. $n(A \cup B) = n(A) + n(B) - n(A \cap B)$
3. $n(A \cup B) = n(A) + n(B)$ (When A and B are disjoint)
4. $n(A \cup B \cup C) = n(A) + n(B) + n(C) - n(A \cap B) - n(B \cap C) - n(C \cap A)$
   $+ n(A \cap B \cap C)$
5. $n(A^c \cup B^c) = n[(A \cap B)^c] = n(U) - n(A \cap B)$
6. $n(A^c \cap B^c) = n[(A \cup B)^c] = n(U) - n(A \cup B)$
7. $n(Exactly\ two\ sets) = n(A \cap B) + n(B \cap C) + n(C \cap A) - 3n(A \cap B \cap C)$
8. $n(Exactly\ one\ set) = n(A) + n(B) + n(C) - 2n(A \cap B) - 2n(B \cap C) - 2n(C \cap A) + 3n(A \cap B \cap C)$

Laws of Algebra of sets:

1. **Idempotent Law:** $A \cup A = A \cap A = A$
2. **Commutative Law:** $A \cup B = B \cup A\ ; A \cap B = B \cap A$
3. **Associative Law:** $A \cup (B \cup C) = (A \cup B) \cup C\ ; A \cap (B \cap C) = (A \cap B) \cap C$
4. **Distributive Law:** $A \cup (B \cap C) = (A \cup B) \cap (A \cup C)\ ;$
   $A \cap (B \cup C) = (A \cap B) \cup (A \cap C)$
5. **Existence of Identity:** $A \cup \emptyset = \emptyset \cup A = A\ ; A \cap U = U \cap A = A$
6. **Complement Laws:** $A \cup A^c = U\ ; A \cap A^c = \emptyset\ ; \emptyset^c = U\ ; U^c = \emptyset$
7. **DeMorgan's Laws:** $(A \cup B)^c = A^c \cap B^c\ ; (A \cap B)^c = A^c \cup B^c\ ; (A^c)^c = A$

7. RELATIONS

Cartesian Product: Let A and B be two non empty sets. Then the cartesian pproduct
is denoted by $A \times B$ and read as A cross B.
It is defined as $A \times B = \{(a,b) \setminus a \in A \text{ and } b \in B\}$
If $n(A) = m$ and $n(B) = n$ then $n(A \times B) = mn$

Properties of Cartesian Product:
- (i) If $A \times B = \emptyset$ then either A or B is an empty set
- (ii) $A \times (B \cup C) = (A \times B) \cup (A \times C)$
- (iii) $A \times (B \cap C) = (A \times B) \cap (A \times C)$
- (iv) $(A \times B) \times (C \times D) = (A \cap C) \times (B \cap D)$
- (v) -1If $A \subseteq B$ and $C \subseteq D$ then $(A \times C) \subseteq (B \times D)$
- (vi) If $A \times B = B \times A$ then $A = B$
- (vii) If A and B have n elements in common then the number of elements common to $A \times B$ and $B \times A$ is n^2

Relation:

Let A and B be two non empty sets. Then any subset of $A \times B$ is called as a relation from A to B and is denoted as $A\mathcal{R}B$.
If $n(A) = m$ and $n(B) = n$ then the number of relations from A to B is 2^{mn}

Types of Relations: Let R be a relation from A to B
1. **Reflexive:** R is a reflexive $\leftrightarrow (a,a) \in R \forall a \in A$
2. **Symmetric:** R is symmetric $\leftrightarrow If (a,b) \in R$ then $(b,a) \in R$
3. **Transitive:** R is transitive $\leftrightarrow If (a,b) \in R$ and $(b,c) \in R$ then $(a,c) \in R$
4. **Equivalence:** If R is a reflexive, symmetric and transitive then R is an equivalence relation.
5. **Anti Symmetric:** R is anti symmetric $\leftrightarrow If (a,b) \in R$ and $(b,a) \in R$ then $a = b$

Some more relations:
1. **Empty (Or) Void Relation:** If $R = \emptyset$ then R is called as an empty relation.
2. **Universal Relation:** If $R = A \times B$ then R is called as an universal relation.
3. **Inverse Relation:** R^{-1} is the inverse relation of R
   i.e. $If (a,b) \in R$ then $(b,a) \in R^{-1}$
4. **Identity Relation:** $R = \{(a,a) \forall a \in A\}$ is an identity relation in A

Note:

1. Number of reflexive relations on a set with n number of elements is $2^{n(n-1)}$
2. Number of symmetric relations on a set with n number of elements is $2^{\frac{n(n+1)}{2}}$
3. Number of relations on a set with n number of elements which are both reflexive and symmetric is $2^{\frac{n(n-1)}{2}}$

8. FUNCTIONS

Function: A relation f from A to B is said to be a function when "for every element $a \in A$ then there exist unique $b \in B$ such that $f(a) = b$". The function f from A to B is denoted by $f: A \to B$.

Here b is called as image of a and a is called as pre-image of b.

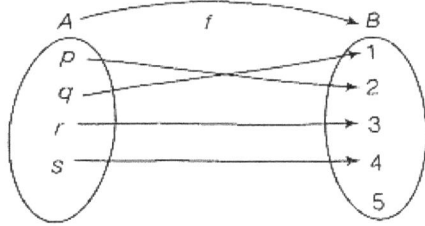

Every function is a relation but each relation need not be a function.

Domain, Co-domain and Range: If $f: A \to B$ is a function then A is called domain, B is called co-domain and set of images is called as a range.

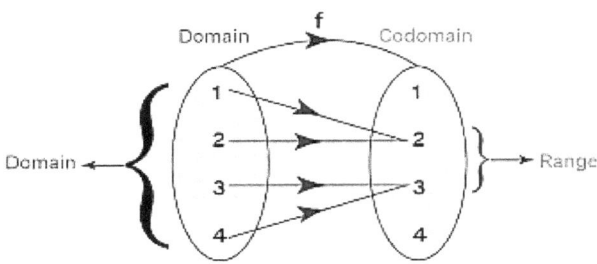

Types of Functions:

1. **One-One Function (or) Injection:** Let $f: A \to B$ be a function. Then f is said to be an one-one function or Injection when distinct elements of A have distinct images in B.

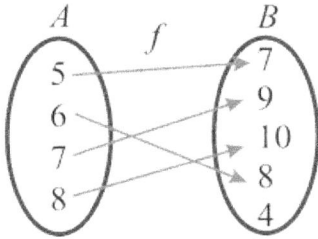

(i) If $f(x_1) = f(x_2)$ then $x_1 = x_2 \ \forall \ x_1, x_2 \in A$
(ii) The necessary condition for one-one function is $n(A) \leq n(B)$

2. **Onto Function (or) Surjection:** Let $f: A \to B$ be a function. Then f is said to be an onto function or Surjection when "for every $b \in B \; \exists$ some $a \in A \; \ni f(a) = b$".

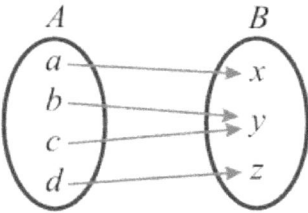

   (i) Range=Co-domain
   (ii) The necessary condition for onto function is $n(A) \geq n(B)$

3. **Bijection:** Let $f: A \to B$ be a function. Then f is said to be bijection when f is both one-one and onto.

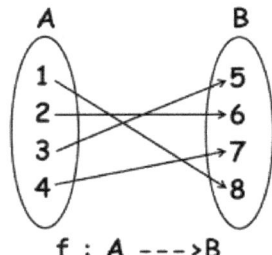

   The necessary condition for bijection is $n(A) = n(B)$

4. **Many one Function:** The function which is not an one-one function is called as a many one function.

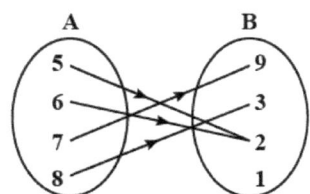

5. **Into Function:** The function which is not onto function is called as a into function

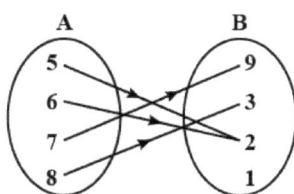

Number of functions:

1. If $n(A) = m$ and $n(B) = n$ then the number of relations from A to B = 2^{mn}
2. If $n(A) = m$ and $n(B) = n$ then the number of functions from A to B = n^m
3. If $n(A) = m$ and $n(B) = n$ then the number of relations which are not functions = $2^{mn} - n^m$
4. If $n(A) = m$ and $n(B) = n$ then the number of one-one functions from A to B = n_{P_m}
5. If $n(A) = m$ and $n(B) = n$ then the number of onto functions
 = $\sum_{r=0}^{n}(-1)^r n_{C_r}(n-r)^m$
6. If $n(A) = n(B) = n$ then the number of bijections = $n!$
7. The number of many one functions = Number of functions – Number of one-one functions
8. The number of into functions = Number of functions – Number of Onto functions
9. The number of non-bijective functions = Number of functions – Number of Bijections

Special types of Functions:

1. **Identity Function:** The function $f: R \to R$ is said to be an identity function when

   $f(x) = x \ \forall x \in R$

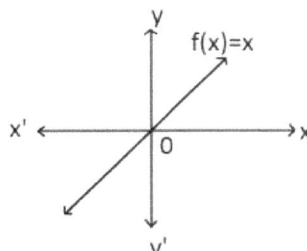

2. **Constant Function:** The function $f: R \to R$ is said to be a constant function when

   $f(x) = k \ \forall x \in R$

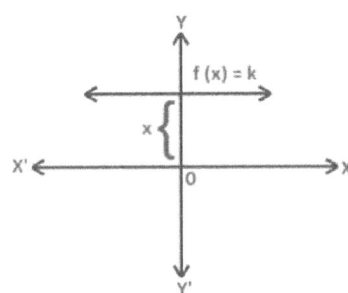

3. **Polynomial Function:** The function $f: R \to R$ is defined by $f(x) = a_0 + a_1 x + a_2 x^2 + \cdots \ldots \ldots a_n x^n$ where $n \in N$ and $a_0, a_1, a_2, \ldots \ldots a_n \in R \ \forall x \in R$ is called as a polynomial function.

4. **Rational Function:** Let $f(x)$ and $g(x)$ be polynomial functions in x then $\frac{f(x)}{g(x)}, g(x) \neq 0$ is called as a rational function.

5. **Modulus Function:** The function $f: R \to R$ is defined by $f(x) = |x| = \begin{cases} x, if \ x \geq 0 \\ -x, if \ x \leq 0 \end{cases} \forall x \in R$ is called as a modulus function.

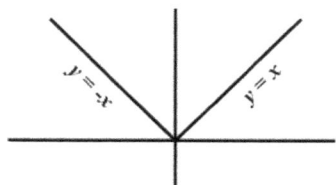

6. **Signum Function:** The function $f: R \to R$ is defined by $f(x) = \begin{cases} 1, if \ x > 0 \\ 0, if \ x = 0 \\ -1, if \ x < 0 \end{cases}$

   is called as a signum function.

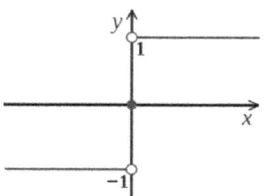

7. **Reciprocal Function:** The function $f: R - \{0\} \to R$ is defined by $f(x) = \frac{1}{x} \ \forall x \in R - \{0\}$ is called as a reciprocal function.

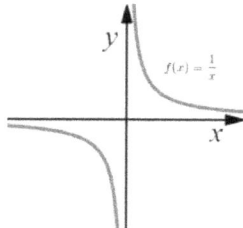

8. **Exponential Function:** The function $f: R \to (0, \infty)$ is defined by $f(x) = a^x$ where $a > 0$ and $a \neq 1 \ \forall x \in R$ is called as an exponential function.

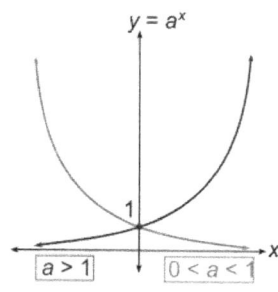

9. **Logarithmic Function:** The function $f: (0, \infty) \to R$ is defined by $f(x) = \log_a x$ where $a > 0$ and $a \neq 1$ $\forall x \in (0, \infty)$ is called as a logarithmic function.

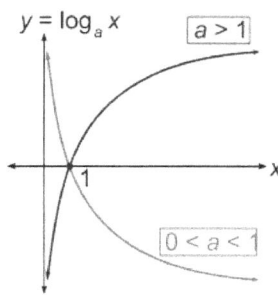

10. **Greatest integer Function (or) Step Function:** The function $f: R \to R$ is defined by $f(x) = [x]$ $\forall x \in R$ takes the value of the greatest integer less than or equal to x.

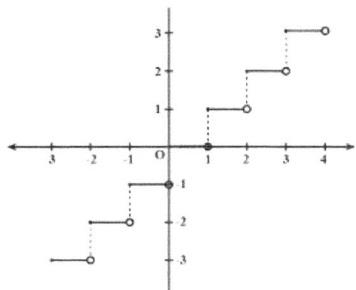

11. **Fractional part Function:** The function $f: R \to R$ is defined by $f(x) = \{x\}$ $= x - [x]$ $\forall x \in R$.

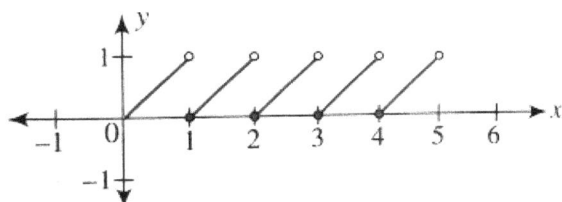

12. **Even and Odd Functions:**
    (i) The function $f(x)$ is said to be an even function when $f(-x) = f(x)$
    Eg: $f(x) = x^4 + x^2 + 5$; $f(x) = \cos x$
    (ii) The function $f(x)$ is said to be an even function when $f(-x) = -f(x)$
    Eg: $f(x) = x^3 + x$; $f(x) = \sin x$

13. **Explicit Function:** A function $y = f(x)$ is said to be an explicit function when the dependent variable y can be expressed in terms of the independent variable x.
    Eg: $y = x^2 + x + 1$

14. Implicit Function: A function $y = f(x)$ is said to be an implicit function when the dependent variable y cannot be expressed in terms of the independent variable x.

Eg: $x^3y + xy - 2x^2 + 5x - 3 = 0$

15. Periodic Function: A function $f(x)$ is said to be a periodic function when

$f(x + P) = f(x) \forall x \in domain\ of\ f$, where P is the least positive integer. Here P is called as period of the function.

Eg: $f(x) = cosx$

Real valued function: A function $f: A \to B$ is called as a real valued function when $B \subseteq R$.

If $A \subseteq R\ and\ B \subseteq R$ then f is called as real function.

Properties of Real valued functions:

Let f and g be two real valued functions with domain A and B respectively. Then both f and g are defined on $A \cap B$ where $A \cap B \neq \emptyset$. Then

(i) $(f \pm g)(x) = f(x) \pm g(x) \forall x \in A \cap B$
(ii) $(fg)(x) = f(x).g(x) \forall x \in A \cap B$
(iii) $\left(\frac{f}{g}\right)(x) = \frac{f(x)}{g(x)} \forall x \in A \cap B$
(iv) $(f \pm k)(x) = f(x) \pm k\ for\ some\ scalar\ k$
(v) $(kf)(x) = k.f(x)\ f for\ some\ scalar\ k$
(vi) $f^n(x) = [f(x)]^n\ \forall n > 0$
(vii) $|f|(x) = |f(x)|$

Equality of functions: The two functions f and g are said to be equal when

(i) The domain of f = The domain of g
(ii) $f(x) = g(x) \forall x \in the\ same\ domain$

Composite Function: If $f: A \to B$ and $g: B \to C$ are two functions then $gof: A \to C$ is defined by $gof(x) = g\{f(x)\} \forall x \in A$ is called the composite function f and g.

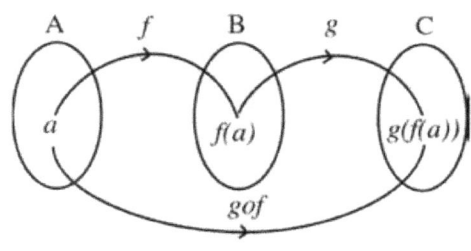

Inverse Function: A function $f: A \to B$ is said to be invertible if there exists a function $g: B \to A$ such that $gof = I_A$ and $fog = I_B$. Then the function g is said to be inverse of f and it is denoted by f^{-1}.

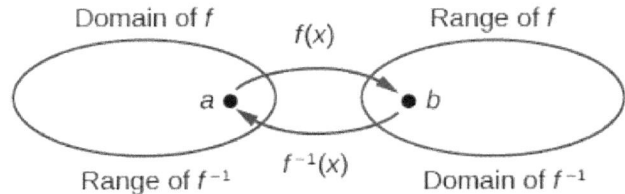

If the function f has inverse then f must be bijective function.

Properties of Composite and Inverse Functions:

(i) In general, $fog \neq gof$
(ii) If $f: A \to B$ and $g: B \to C$ are one-one then $gof: A \to C$ is also one-one
(iii) If $f: A \to B$ and $g: B \to C$ are onto then $gof: A \to C$ is also onto
(iv) If $f: A \to B$ and $g: B \to C$ are bijections then $gof: A \to C$ is also bijection
(v) If $f: A \to B$ and $g: B \to C$ are bijections then $(gof)^{-1} = f^{-1}og^{-1}$
(vi) If $f: A \to B$ is a function then $foI_A = I_Bof = f$
(vii) If $f: A \to B$ is a bijection then $fof^{-1} = I_B$ and $f^{-1}of = I_A$
(viii) If $f: A \to B$ and $g: B \to C$ are functions such that $gof = I_A$ and $fog = I_B$ then $g = f^{-1}$
(ix) If $gof: A \to C$ is one-one then f is one-one
(x) If $gof: A \to C$ is onto then g is onto
(xi) If If $f: A \to B$, $g: B \to C$ and $h: C \to D$ are functions then $ho(gof) = (hog)of$

Some more important functions:

1. If $f(x + y) = f(x) + f(y)$ then $f(x) = kx$ for some scalar k
2. If $f(x + y) = f(x).f(y)$ then $f(x) = k^x$ for some scalar k
3. If $f(xy) = f(x) + f(y)$ then $f(x) = k \log_a x$ for some scalar k and $a > 0$ and $a \neq 1$
4. If $f(x) + f\left(\frac{1}{x}\right) = f(x)f\left(\frac{1}{x}\right)$ then $f(x) = x^n \pm 1$ or $1 \pm x^n$
5. If $f(x + y) + f(x - y) = 2f(x)f(y)$ then $f(x) = \frac{k^x + k^{-x}}{2}$ or $f(x) = \cos kx$

9. MATHEMATICAL REASONING

Statement: A sentence in mathematical approach which is either true or false but not both is called as a statement.

Simple Statement: Any statement whose truth value does not depend on another statement is called as a simple statement.

Compound Statement: A statement which is formed with two or more simple statements by using connectives 'and', 'or', 'implies', 'if and only if' is called as a compound statement.

Truth table: A table shows that relationship between the truth values of compound statements and the truth values of simple statements is called as the truth table.

Conjunction: Let p and q be two simple statements. Then conjunction of p and q is denoted by $p \wedge q$ and read as p and q.

p	q	p∧q
T	T	T
T	F	F
F	T	F
F	F	F

Disjunction: Let p and q be two simple statements. Then disjunction of p and q is denoted by $p \vee q$ and read as p or q.

P	Q	P v Q
T	T	T
T	F	T
F	T	T
F	F	F

Conditional (or) Implication: Let p and q be two simple statements. Then conjunction of p and q is denoted by $p \rightarrow q$ and read as p implies q.

p	q	p→q
T	T	T
T	F	F
F	T	T
F	F	T

Bi implication: Let p and q be two simple statements. Then conjunction of p and q is denoted by $p \leftrightarrow q$ and read as p if and only if q.

p	q	p↔q
T	T	T
T	F	F
F	T	F
F	F	T

Negation: A contradict statement of a given statement is called as a negation.

p	~p
T	F
F	T

Converse, Inverse and Contrapositive: Let p and q be two simple statements. Then

(i) The converse of $p \to q$ is $q \to p$

(ii) The inverse of $p \to q$ is $\sim p \to \sim q$

(iii) The contrapositive of $p \to q$ is $\sim q \to \sim p$

p	q	$p \to q$	$q \to p$ (converse)	$\sim p \to \sim q$ (inverse)	$\sim q \to \sim p$ (contrapositive)
T	T	T	T	T	T
T	F	F	T	T	F
F	T	T	F	F	T
F	F	T	T	T	T

Tautology: If a compound statement is always true then it is called as a tautology.

Contradiction: If a compound statement is always false then it is called as a contradiction.

Logical Equivalence: Two or more compound statements are said to be logical equivalent when they have same entries in the last column.

Eg: $p \to (q \to r)$ is logical equivalent to $(p \to q) \to r$

P	Q	R	$Q \to R$	$P \to Q$	$P \to (Q \to R)$	$(P \to Q) \to R$
T	T	T	T	T	T	T
T	T	F	F	T	F	F
T	F	T	T	F	T	T
T	F	F	T	F	T	T
F	T	T	T	T	T	T
F	T	F	F	T	T	F
F	F	T	T	T	T	T
F	F	F	T	T	T	F

10. SEQUENCES AND SERIES

Sequence: A list of things (usually numbers) is called as a sequence.

   Eg: 1,3,5,7,………

Series: The sum of the terms of a sequence is called as a series.

   Eg: 1+3+5+7+…….. upto n terms

Progression: Sequences which are following a specific pattern are called as a progression.

Types of Progressions: Majorly, four types of progressions are there. They are

   (i) Arithmatic Progression (A.P.)

   (ii) Geometric progression (G.P.)

   (iii) Harmonic Progression (H.P.)

   (iv) Arithmetico geometric Progression (A.G.P.)

Arithmatic Progression (A.P.) : A progression is called as a arithmetic progression when the difference between any two consecutive terms of the progression is same.

   In general, the A.P. is in the form a,a+d,a+2d,a+3d,……

   Here a is the first term and common difference is d.

General term in A.P.: If a is the first term and d is the common difference then n^{th} term of A.P. is denoted as general term. It is denoted by $T_n = a + (n-1)d$

   Clearly, $T_n - T_{n-1} = d$

   T_r from the end $= T_{n-r+1}$ from the beginning where n is the number of terms in A.P.

Sum of n terms of an A.P.:

   (i) If a is the first term, d is the common difference and n is the number of terms then sum of the n terms of the A.P. is $S_n = \frac{n}{2}[2a + (n-1)d]$

   (ii) If a is the first term, l is the last term and n is the number of terms then sum of the n terms of the A.P. is $S_n = \frac{n}{2}(a + l)$

The nth term of A.P. is given by $T_n = S_n - S_{n-1}$

Selection of term in an A.P.:

(i) If an A.P. consists of 3 terms then they can be taken as a-d, a, a+d with common difference d.

(ii) If an A.P. consists of 4 terms then they can be taken as a-3d, a-d, a+d, a+3d

with common difference 2d.

(ii) If an A.P. consists of 5 terms then they can be taken as a-2d, a-d, a, a+d, a+2d
with common difference d.

Properties of A.P.:

(i) If a, b, c are in A.P. then 2b=a+c

(ii) If $a_1, a_2, a_3, \ldots a_n$ are in A.P. then $a_1 + k, a_2 + k, a_3 + k, \ldots a_n + k$ are also in A.P. for some scalar k

(iii) If $a_1, a_2, a_3, \ldots a_n$ are in A.P. then $ka_1, ka_2, ka_3, \ldots ka_n$ are also in A.P. for some scalar k

Geometric Progression (G.P.): A progression is called as a geometric progression when the ratio of any two consecutive terms of the progression is same.

In general, the G.P. is in the form $a, ar, ar^2, ar^3, \ldots\ldots\ldots$

Here a is the first term and common ratio is r.

General term in G.P.: If a is the first term and r is the common ratio then n^{th} term of G.P. is denoted as general term. It is denoted by $T_n = ar^{n-1}$

Clearly, $\frac{T_n}{T_{n-1}} = r$

T_r from the end = T_{n-r+1} from the beginning where n is the number of terms in G.P.

Sum of n terms of an G.P.: If a is the first term and r is the common ratio then the sum of n terms of the G.P. is given by $S_n = \begin{cases} \frac{a(r^n-1)}{r-1} & when\ r > 1 \\ \frac{a(1-r^n)}{1-r} & when\ r < 1 \end{cases}$

Sum of infinite terms of an G.P.: If a is the first term and r is the common ratio then the sum of infinite terms of the G.P. is given by $S_\infty = \frac{a}{1-r}$ when $|r| < 1$

Selection of term in an G.P.:

(i) If an G.P. consists 3 terms then they can be taken as $\frac{a}{r}, a, ar$ with common ratio r.

(ii) If an G.P. consists 4 terms then they can be taken as $\frac{a}{r^3}, \frac{a}{r}, ar, ar^3$ with common ratio r^2.

(iii) If an G.P. consists 5 terms then they can be taken as $\frac{a}{r^2}, \frac{a}{r}, a, ar, ar^2$ with common ratio r.

Properties of G.P.:

(i) If a, b, c are in G.P. then $b^2 = ac$

(ii) If $a_1, a_2, a_3, \ldots a_n$ are in G.P. then $\frac{1}{a_1}, \frac{1}{a_2}, \frac{1}{a_3}, \ldots \ldots \frac{1}{a_n}$ are also in G.P. for some scalar k

(iii) If $a_1, a_2, a_3, \ldots a_n$ are in G.P. then $ka_1, ka_2, ka_3, \ldots ka_n$ are also in G.P. for some scalar k

(iv) If $a_1, a_2, a_3, \ldots a_n$ are in G.P. then $a_1^k, a_2^k, a_3^k, \ldots \ldots a_n^k$ are also in G.P. for some scalar k

Arithmetico-geometric Progression (A.G.P.): The progression is in the form of $a, (a+d)r, (a+2d)r^2, (a+3d)r^3, \ldots \ldots \ldots$ is called as a arithmetic geometric progression.

General term in A.G.P.: The general term of A.G.P. is denoted by $T_n = [a + (n-1)d]r^{n-1}$ where a is the first term, d is the common difference and r is the common ratio.

Sum of n terms of an A.G.P.: If a is the first term, d is the common difference and r is the common ratio then the sum of n terms of the A.G.P. is given by

$$S_n = \begin{cases} \frac{a}{1-r} + dr\frac{(1-r^{n-1})}{(1-r)^2} - \frac{[a+(n-1)d]r^n}{1-r} \text{ when } r \neq 1 \\ \frac{n}{2}[2a+(n-1)d] \text{ when } r = 1 \end{cases}$$

Sum of infinite terms of an A.G.P.: If a is the first term, d is the common difference and r is the common ratio then the sum of n terms of the A.G.P. is given by $S_\infty = \frac{a}{1-r} + \frac{dr}{(1-r)^2}$ when $|r| < 1$

Harmonic Progression (H.P.): The reciprocal progression of A.P. is called as a H.P. i.e. the H.P. is in the form of $\frac{1}{a}, \frac{1}{a+d}, \frac{1}{a+2d}, \ldots \ldots \frac{1}{a+(n-1)d}$ where first term is $\frac{1}{a}$ and general term, $T_n = \frac{1}{a+(n-1)d}$

Properties of H.P.:

(i) If a, b, c are in H.P. then $b = \frac{2ac}{a+c}$

(ii) If $a_1, a_2, a_3, \ldots a_n$ are in H.P. then $\frac{1}{a_1}, \frac{1}{a_2}, \frac{1}{a_3}, \ldots \ldots \frac{1}{a_n}$ are in A.P. for some scalar k

(iii) If $a_1, a_2, a_3, \ldots a_n$ are in H.P. then $ka_1, ka_2, ka_3, \ldots ka_n$ are also in H.P. for some scalar k

Relation between A.M., G.M., H.M. :

Let a, b be two real numbers and A,G,H are the Arithmetic mean (A.M.), Geometric mean (G.M.), Harmonic mean (H.M.) respectively then

$$A = \frac{a+b}{2}, G = \sqrt{ab}, H = \frac{2ab}{a+b}$$

Clearly, $A.M. \geq G.M. \geq H.M.$

Properties of A.M., G.M., H.M. :

(i) The arithmetic mean of $a_1, a_2, a_3, \ldots a_n$ is $A = \frac{a_1 + a_2 + a_3 + \ldots + a_n}{n}$

(ii) The geometric mean of $a_1, a_2, a_3, \ldots a_n$ is $G = (a_1 \cdot a_2 \cdot a_3 \cdot \ldots \cdot a_n)^{1\backslash n}$

(iii) The harmonic mean of $a_1, a_2, a_3, \ldots a_n$ is H then $\frac{1}{H} = \frac{1}{n}\left(\frac{1}{a_1} + \frac{1}{a_2} + \frac{1}{a_3} + \cdots \cdot \frac{1}{a_n}\right)$

Weighted means: Let $a_1, a_2, a_3, \ldots a_n$ be n positive real numbers and $m_1, m_2, m_3, \ldots m_n$ be n positive rational numbers. Then we describe weighted Arithmetic mean (A), Geometric mean (G), Harmonic mean (H) as follows

(i) $A = \frac{m_1 a_1 + m_2 a_2 + m_3 a_3 + \cdots + m_n a_n}{m_1 + m_2 + m_3 + \ldots + m_n}$

(ii) $G = (a_1^{m_1} \cdot a_2^{m_2} \cdot a_3^{m_3} \ldots \ldots a_n^{m_n})^{\frac{1}{m_1 + m_2 + m_3 + \ldots + m_n}}$

(iii) $H = \frac{m_1 + m_2 + m_3 + \ldots + m_n}{\frac{m_1}{a_1} + \frac{m_2}{a_2} + \frac{m_3}{a_3} + \cdots + \frac{m_n}{a_n}}$

11. MATRICES

Matrix: The rectangular array of entries (either numbers or letters) is called as a matrix.

They are denoted by the brackets [ ] or ( ).

The entries are called elements.

Generally, the matrices are represented by the capital letters A, B, C,……. And elements are represented by small letters a, b,c,….

The elements entered in horizontal are called as row elements and entered in vertical are called as column elements.

Order of a matrix: If a matrix has m rows and n columns then the order is represented by $m \times n$ and read as m by n.

Representation of a matrix: A matrix $A = [a_{ij}]$ of order $m \times n$ is represented by the elements such that a_{ij} is the element present in i^{th} row and j^{th} column.

i.e. $A = \begin{bmatrix} a_{11} & a_{12} & a_{13} & \cdots & a_{1n} \\ \vdots & & & \ddots & \vdots \\ a_{m1} & a_{m2} & a_{m3} & \cdots & a_{mn} \end{bmatrix}$

Types of Matrices

1. Row Matrix: If a matrix has only one row then the matrix is called as a row matrix.

Eg: $[3 \quad 2 \quad -7]$

2. Column Matrix: If a matrix has only one column then the matrix is called as a column matrix.

Eg: $\begin{bmatrix} 1 \\ 0 \\ 5 \end{bmatrix}$

3. Rectangular Matrix: If the number of rows is not equal to the number of columns of a matrix then the matrix is called as a rectangular matrix.

Eg: $\begin{bmatrix} 2 & 0 & -3 \\ 0 & 5 & 9 \end{bmatrix}$ has order 2×3

4. Square matrix: If the number of rows is equal to the number of columns of a matrix then the matrix is called as a square matrix.

Eg: $\begin{bmatrix} 1 & -5 & 0 \\ 4 & 6 & -2 \\ -3 & 0 & 5 \end{bmatrix}$

5. Zero Matrix (or) Null Matrix : If all the elements of a matrix are zero then the matrix is called as a zero matrix. It is denoted by O.

Eg: $O = \begin{bmatrix} 0 & 0 \\ 0 & 0 \\ 0 & 0 \end{bmatrix}$

6. Principal diagonal: In a square matrix, the elements present in a_{ij} where $i = j$ are called as principal diagonal elements.

7. Trace of a square matrix: The sum of the principal diagonal elements of a square matrix is called as trace of the matrix.

Eg: The trace of $\begin{bmatrix} 1 & 0 & 2 \\ 3 & -1 & 4 \\ -5 & 6 & 1 \end{bmatrix} = 1 - 1 + 1 = 1$

Properties of trace: If A, B and C are two square matrices of same order then

(i) $t_r(kA) = kt_r(A)$ for some scalar k

(ii) $t_r(A \pm B) = t_r(A) \pm t_r(B)$

(iii) $t_r(AB) = t_r(BA)$

(iv) $t_r(AB) \neq t_r(A).t_r(B)$

(v) $t_r(I_n) = n$

(vi) $t_r(ABC) = t_r(BCA) = t_r(CAB)$

8. Diagonal Matrix: In a square matrix, if the principal diagonal elements are not all zero and the remaining elements all are zero then the matrix is called as a diagonal matrix.

Eg: $\begin{bmatrix} 1 & 0 & 0 \\ 0 & 5 & 0 \\ 0 & 0 & -3 \end{bmatrix}$

9. Scalar Matrix: In a square matrix, if the principal diagonal elements all are equal to some scalar except 0 and 1 and the remaining elements all are zero then the matrix is called as a scalar matrix.

Eg: $\begin{bmatrix} 5 & 0 & 0 \\ 0 & 5 & 0 \\ 0 & 0 & 5 \end{bmatrix}$

10. Identity Matrix (or) Unit Matrix: In a square matrix, if the principal diagonal elements all are equal to 1 and the remaining elements all are zero then the matrix is called as an identity matrix.

It is denoted by I.

Eg: $I_3 = \begin{bmatrix} 1 & 0 & 0 \\ 0 & 1 & 0 \\ 0 & 0 & 1 \end{bmatrix}; I_2 = \begin{bmatrix} 1 & 0 \\ 0 & 1 \end{bmatrix}$

11. Triangular Matrices:

(i) Upper triangular matrix: A square matrix $A = [a_{ij}]$ is called as an upper triangular matrix when $a_{ij} = 0 \; for \; i > j$

Eg: $\begin{bmatrix} 1 & 2 & 3 \\ 0 & 4 & 5 \\ 0 & 0 & 6 \end{bmatrix}$

(ii) Lower triangular matrix: A square matrix $A = [a_{ij}]$ is called as a lower triangular matrix when $a_{ij} = 0 \; for \; i < j$

Eg: $\begin{bmatrix} 1 & 0 & 0 \\ 2 & 3 & 0 \\ 4 & 5 & 6 \end{bmatrix}$

Equality of Matrices: Two matrices A and B are said to be equal when they have same order and corresponding elements of the two matrices are equal.

Eg: If $\begin{bmatrix} a & b & c \\ d & e & f \end{bmatrix} = \begin{bmatrix} 1 & 3 & 5 \\ 0 & -2 & 4 \end{bmatrix}$ then

$a = 1, b = 3, c = 5, d = 0, e = -2, f = 4$

Additive inverse of a Matrix: Let A be a matrix then the additive inverse of A is denoted by -A and formed by each element of A is multiplied by negative.

Eg: If $A = \begin{bmatrix} 1 & 0 & -3 \\ -5 & 2 & 7 \end{bmatrix}$ then $-A = \begin{bmatrix} -1 & 0 & 3 \\ 5 & -2 & -7 \end{bmatrix}$

Addition of matrices: The addition of matrices is possible only when they have same order.

Let A and B be two matrices. Then the addition of the matrices A+B is formed by adding corresponding elements of the two matrices.

i.e. if $A = [a_{ij}]_{m \times n}$ and $B = [b_{ij}]_{m \times n}$ then $A + B = [c_{ij}]_{m \times n}$ where $c_{ij} = a_{ij} + b_{ij}$

Properties of Addition: Let A, B of same order and O be the zero matrix of the same order then

   (i) Commutative Law: $A + B = B + A$

   (ii) Associative Law: $A + (B + C) = (A + B) + C$

   (iii) Existence of Identity: $A + O = O + A = A$

   (iv) Additive inverse: $A + (-A) = (-A) + A = O$

Multiplication of a matrix by a scalar: Let A be a matrix and k be a scalar then the scalar multiplication of A with k is kA is obtained when each element of A is multiplied by k.

Eg: If $A = \begin{bmatrix} 1 & 2 & 3 \\ 0 & 2 & 4 \end{bmatrix}$ then $3A = \begin{bmatrix} 3 & 6 & 9 \\ 0 & 6 & 12 \end{bmatrix}$

Properties: Let A, B be two matrices of same order and k, l be two scalars then

(i) $k(A + B) = kA + kB$ **(ii)** $(k + l)A = kA + lA$ **(iii)** $k(lA) = l(kA) = kl(A)$

Multiplication of matrices: Let A and B be two matrices then the multiplication AB is possible when the number of columns in A is equal to the number of rows in B.

If $A = [a_{ij}]_{m \times n}$ and $B = [b_{jk}]_{n \times p}$ then $AB = [c_{ij}]_{m \times p}$ where $c_{ij} = \sum_{j=1}^{n}(a_{ij} \cdot b_{jk})$

Properties of Multiplication: Let A, B, C be three matrices such that their multiplication is possible then

    (i) In general, $AB \neq BA$

    (ii) $A(BC) = (AB)C$

    (iii) $A(B + C) = AB + AC$

    (iv) If $AB = O$ then either A or B need not be equal to zero matrix

    (v) $AI = IA = A$ where A is a square matrix of order of I

    (vi) $(A^m)^n =) (A^n)^m = A^{mn}$

    (vii) $A^m \cdot A^n = A^{m+n}$

Transpose of a matrix: Let A be matrix then the transpose matrix of A is denoted by A^T or A' and is formed by the changing of rows as columns and columns as rows.

If the order of A is $m \times n$ then the order of A^T is $n \times m$

    Eg: If $A = \begin{bmatrix} 1 & 2 & 0 \\ 4 & -3 & 2 \end{bmatrix}$ then $A^T = \begin{bmatrix} 1 & 4 \\ 2 & -3 \\ 0 & 2 \end{bmatrix}$

Properties of Transpose: Let A, B be two matrices and k be a scalar then

(i) $(A^T)^T = A$ **(ii)** $(A \pm B)^T = A^T \pm B^T$ **(iii)** $(AB)^T = B^T \cdot A^T$

(iv) $(kA)^T = kA^T$

Symmetric Matrix: Let A be a square matrix. Then A is said to be a symmetric matrix when $A^T = A$

    Eg: $A = \begin{bmatrix} 1 & 2 & 3 \\ 2 & 4 & 5 \\ 3 & 5 & 6 \end{bmatrix}$ is a symmetric matrix since $A^T = A$

Skew-Symmetric Matrix: Let A be a square matrix. Then A is said to be a symmetric matrix when $A^T = -A$

    Eg: $A = \begin{bmatrix} 0 & 1 & 2 \\ -1 & 0 & -3 \\ -2 & 3 & 0 \end{bmatrix}$ is a skew-symmetric matrix since $A^T = -A$

Let A be a square matrix then A can be uniquely expressed as a sum of a symmetric matrix and skew-symmetric matrix.

i.e. $A = \frac{A+A^T}{2} + \frac{A-A^T}{2}$ where $\frac{A+A^T}{2}$ is symmetric matrix and $\frac{A-A^T}{2}$ is a skew-symmetric matrix.

Some more special matrices:

(i) **Orthogonal Matrix:** A square matrix A is said to be orthogonal matrix When $AA^T = A^T A = I$

(ii) **Idempotent Matrix:** A square matrix A is said to be idempotent matrix when $A^2 = A$

(iii) **Involutary Matrix:** A square matrix A is said to be involutary matrix when $A^2 = I$

(iv) **Nilpotent Matrix:** A square matrix A is said to be nilpotent matrix when $A^n = O$ for some $n \in N$

(v) **Conjugate of a Matrix:** Let A be a matrix. Then the conjugate matrix of A is denoted by $\bar{A}$ and is obtained by replacing the elements by their conjugates.

The transpose of conjugate matrix is denoted by A^θ

(vi) **Hermitian Matrix:** A square matrix A is said to be Hermitian when $A^\theta = A$

(vii) **Skew-Hermitian Matrix:** A square matrix A is said to be Skew-Hermitian when $A^\theta = -A$

(viii) **Unitary Matrix:** A square matrix A is said to be unitary matrix when $A^\theta A = AA^\theta = I$

12. DETERMINANTS

Determinant of a 2 × 2 Matrix: If $A = \begin{bmatrix} a & b \\ c & d \end{bmatrix}$ then the determinant of A is denoted by detA or $|A|$ and is defined by $detA = |A| = \begin{vmatrix} a & b \\ c & d \end{vmatrix} = ad - bc$.

Minor of an element in 3 × 3 matrix: The determinant of the 2 × 2 matrix which formed after deleting the row and column in which the element present is called as minor of the element of a 3 × 3 matrix.

Let $A = [a_{ij}]$ be a square matrix. Then the minor of an element a_{ij} is the determinant of the 2 × 2 matrix which formed after deleting the ith row and j^{th} column.

Co factor of an element in 3 × 3 matrix: Let $A = [a_{ij}]$ be a square matrix. Then the cofactor of an element a_{ij} is the product of minor of a_{ij} and $(-1)^{i+j}$.

If $A = \begin{bmatrix} a_1 & b_1 & c_1 \\ a_2 & b_2 & c_2 \\ a_3 & b_3 & c_3 \end{bmatrix}$ then cofactor matrix is given by $\begin{bmatrix} A_1 & B_1 & C_1 \\ A_2 & B_2 & C_2 \\ A_3 & B_3 & C_3 \end{bmatrix}$

Determinant of a 3 × 3 matrix: The determinant of 3 × 3 matrix is the sum of the product of the elements with their cofactors of a row or a column. It is denoted by Δ.

i.e. If $A = \begin{bmatrix} a_1 & b_1 & c_1 \\ a_2 & b_2 & c_2 \\ a_3 & b_3 & c_3 \end{bmatrix}$ and cofactor matrix of A is $\begin{bmatrix} A_1 & B_1 & C_1 \\ A_2 & B_2 & C_2 \\ A_3 & B_3 & C_3 \end{bmatrix}$ then determinant of A is given by $\Delta = a_1 A_1 + b_1 B_1 + c_1 C_1 = a_1 A_1 + a_2 A_2 + a_3 A_3 = \cdots \ldots \ldots$

Properties of determinants:

(i) If the two rows or two columns are changed then the sign of the determinant value is changed.

Eg: $\begin{vmatrix} 1 & 2 & 3 \\ 0 & -2 & 4 \\ 3 & 5 & -3 \end{vmatrix} = - \begin{vmatrix} 1 & 3 & 2 \\ 0 & 4 & -2 \\ 3 & -3 & 5 \end{vmatrix}$

(ii) If all the elements of two rows are two columns equal then the value of determinant is zero.

Eg: $\begin{vmatrix} 1 & 2 & 3 \\ 1 & 2 & 3 \\ 4 & 5 & 6 \end{vmatrix} = 0$

(iii) If all the elements of a row or a column are in the same ratio with all the elements of another row or a column then the value of determinant is zero.

$$\text{Eg: :} \begin{vmatrix} 1 & 2 & 3 \\ 2 & 4 & 6 \\ 4 & 5 & 6 \end{vmatrix} = 0$$

(iv) If all the elements of a row or a column are the sum of two terms then the determinant of the matrix can be expressed as the sum of two determinants.

$$\text{Eg: } \begin{vmatrix} a_1+x_1 & b_1 & c_1 \\ a_2+x_2 & b_2 & c_2 \\ a_3+x_3 & b_3 & c_3 \end{vmatrix} = \begin{vmatrix} a_1 & b_1 & c_1 \\ a_2 & b_2 & c_2 \\ a_3 & b_3 & c_3 \end{vmatrix} + \begin{vmatrix} x_1 & b_1 & c_1 \\ x_2 & b_2 & c_2 \\ x_3 & b_3 & c_3 \end{vmatrix}$$

(v) If all the elements of a row or a column are multiplied by k then the value of determinant is also multiplied by k.

$$\text{Eg: } \begin{vmatrix} ka_1 & b_1 & c_1 \\ ka_2 & b_2 & c_2 \\ ka_3 & b_3 & c_3 \end{vmatrix} = k \begin{vmatrix} a_1 & b_1 & c_1 \\ a_2 & b_2 & c_2 \\ a_3 & b_3 & c_3 \end{vmatrix}$$

(vi) $det(kA) = k^3 \, detA$

(vii) $det(AB) = detA.detB = detB.detA$

(viii) $det(A^T) = detA$

(ix) $det(A^n) = (detA)^n$

(x) The sum of the products of the elements of a row or a column with the cofactors of another row or a column is zero.

If $A = \begin{bmatrix} a_1 & b_1 & c_1 \\ a_2 & b_2 & c_2 \\ a_3 & b_3 & c_3 \end{bmatrix}$ and cofactor matrix of A is $\begin{bmatrix} A_1 & B_1 & C_1 \\ A_2 & B_2 & C_2 \\ A_3 & B_3 & C_3 \end{bmatrix}$ then

$a_1A_2 + b_1B_2 + c_1C_2 = 0$

(xi) The determinant of the triangular matrix is the product of their principal diagonal elements.

Eg: The det of $\begin{bmatrix} 1 & 2 & 3 \\ 0 & 4 & 5 \\ 0 & 0 & 6 \end{bmatrix}$ is 24 which is the product of principal diagonal elements.

(xii) If all the elements of a square matrix are polynomials in x and the value determinant is zero when $x = a$ then $x - a$ is a factor of the determinant.

Singular and non-singular Matrices:

The square matrix is said to be singular when the determinant value of the matrix is zero.

The square matrix is said to be non-singular when the determinant value of the matrix is not a zero.

Adjoint of a Matrix: The transpose of a cofactor matrix is called as an adjoint matrix.

The adjoint of $\begin{bmatrix} a & b \\ c & d \end{bmatrix}$ is denoted by $\begin{bmatrix} d & -b \\ -c & a \end{bmatrix}$

If $A = \begin{bmatrix} a_1 & b_1 & c_1 \\ a_2 & b_2 & c_2 \\ a_3 & b_3 & c_3 \end{bmatrix}$ and cofactor matrix of A is $\begin{bmatrix} A_1 & B_1 & C_1 \\ A_2 & B_2 & C_2 \\ A_3 & B_3 & C_3 \end{bmatrix}$ then

$Adj\ A = \begin{bmatrix} A_1 & A_2 & A_3 \\ B_1 & B_2 & B_3 \\ C_1 & C_2 & C_3 \end{bmatrix}$

Properties of Adjoint matrix: If A, B are square matrices of order n and I b the identity matrix of order n then

(i) $A(Adj\ A) = (Adj\ A)A = detA.I$

(ii) $\det(Adj\ A) = (\det A)^{n-1} = \det(cofactor\ of\ A)$

(iii) $Adj(Adj\ A) = (\det A)^{n-2} A$

(iv) $\det(Adj(Adj\ A)) = (\det A)^{(n-1)^2}$

(v) $\det\left(Adj(Adj(Adj\ ....\ m\ times\ (Adj\ A))\right) = (deta)^{(n-1)^m}$

(vi) $Adj(AB) = (Adj\ A)(Adj\ B)$

(vii) $Adj(A^m) = (Adj\ A)^m$

(viii) $Adj(A^T) = (Adj\ A)^T$

(ix) $Adj(kA) = k^{n-1}\ Adj\ A$ for some scalar k.

Invertible: A square matrix A is said to be invertible then there exists another square matrix B of same order such that $AB = BA = I$

Inverse of a square matrix: The inverse of a square matrix A is denoted by A^{-1} and is defined as $A^{-1} = \frac{1}{\det A} Adj\ A$

Properties of Inverse:

(i) $(A^{-1})^{-1} = A$ (ii) $(AB)^{-1} = B^{-1}A^{-1}$ (iii) $(A^T)^{-1} = (A^{-1})^T$

(iv) $\det(A^{-1}) = (detA)^{-1}$ (v) $AA^{-1} = A^{-1}A = I$

(vi) $(A^k)^{-1} = (A^{-1})^k$ for some scalar k

Rank of a matrix: Let A be a non-zero matrix. Then the rank of the matrix is defined as the number of non zero rows of maximum possible square sub matrices after getting an upper triangular matrix by using elementary row operations.

Non-Homogeneous system of 3 equations: The system of equations represented by $AX = B$ is called as a non-homogeneous system of equations.

Consistent and inconsistent systems and their solutions:

(i) If $Rank(A) = Rank(A\ B) = 3$

then the system of equations is consistent, and it has unique solution.

The unique solution is $x = \frac{\Delta_1}{\Delta}, y = \frac{\Delta_2}{\Delta}, z = \frac{\Delta_3}{\Delta}$

(ii) If $Rank(A) = Rank(A\ B) < 3$

then the system of equations is consistent and it has infinitely many solutions.

(iii) If $Rank(A) \neq Rank(A\ B)$

then the system of equations is inconsistent and it has no solution.

In other words,

(i) If $\Delta \neq 0$ then the system of equations is consistent and it has unique solution.

The unique solution is $x = \frac{\Delta_1}{\Delta}, y = \frac{\Delta_2}{\Delta}, z = \frac{\Delta_3}{\Delta}$

(ii) If $\Delta = 0\ and\ \Delta_1 = \Delta_2 = \Delta_3 = 0$ then the system of equations is consistent and it has infinitely many solutions.

(iii) If $\Delta = 0$ and at least one of $\Delta_1, \Delta_2, \Delta_3$ is not equal to zero then the system of equations is inconsistent and it has no solution.

Homogeneous system of 3 equations: The system of equations represented by $AX = O$ is called as a homogeneous system of equations.

Trivial and non-trivial solutions:

(i) If $\Delta \neq 0$ then it has trivial solution. The trivial solution is unique solution is given by $x = y = z = 0$

(ii) If $\Delta = 0$ then it has non-trivial solution. It has infinitely many solutions.

Characteristic equation of a matrix: If A is a square matrix of order n and I is the identity matrix of same order then $|A - \lambda I| = 0$ is called as characteristic equation of A.

13. QUADRATIC EQUATIONS AND EXPRESSIONS

Quadratic expression: The polynomial is in the form of $ax^2 + bx + c$ where a,b,c are real or complex numbers and $a \neq 0$ is called as a quadratic expression.

Eg: $2x^2 - 5x + 4, 4x^2 + 9$

Zero of a quadratic expression: Let $f(x)$ be a quadratic expression. Then α is said to be a zero of the expression when $f(\alpha) = 0$.

Eg: 2 and 3 are zeros of $x^2 - 5x + 6$

Discriminant of a quadratic expression: The discriminant of the quadratic expression $ax^2 + bx + c$ is given by $b^2 - 4ac$. It is denoted by Δ.

Quadratic equation: The equation is in the form of $ax^2 + bx + c = 0$ where a,b,c are real or complex numbers and $a \neq 0$ is called as a quadratic expression.

Eg: $x^2 - 5x + 7 = 0, x^2 + 9 = 0$

Root or Solution of a quadratic equation: Let $f(x) = 0$ be a quadratic equation. Then α is said to be a root or solution of the equation when $f(\alpha) = 0$.

Eg: 1 and 3 are roots of $x^2 - 4x + 3 = 0$

Roots of quadratic equation: The roots of a quadratic equation $ax^2 + bx + c = 0$ are given by $x = \frac{-b \pm \sqrt{b^2 - 4ac}}{2a}$

Relation between roots and the coefficients: Let α and β be the roots of $ax^2 + bx + c = 0$ then $\alpha + \beta = \frac{-b}{a}$ and $\alpha\beta = \frac{c}{a}$

Quadratic equation whose roots given: The quadratic equation whose roots are α and β is $(x - \alpha)(x - \beta) = 0$ or $x^2 - (\alpha + \beta)x + \alpha\beta = 0$

Discriminant of a quadratic equation: The discriminant of the quadratic equation $ax^2 + bx + c = 0$ is given by $b^2 - 4ac$. It is denoted by Δ.

Nature of the roots: Let $ax^2 + bx + c = 0$ be a quadratic equation.

   (i) If $\Delta \geq 0$ then the roots are real

   (ii) If $\Delta = 0$ then the roots are real and equal

   (iii) If $\Delta > 0$ and it is a perfect square then the roots are rational and distinct

   (iv) If $\Delta > 0$ and it is not a perfect square then the roots are irrational and conjugate

   (v) If $\Delta < 0$ and it is not a perfect square then the roots are imaginary and conjugate.

Properties of roots:

Let $ax^2 + bx + c = 0$ be a quadratic equation.

(i) If a and c are of same sign then the roots have same sign

(ii) If a and c are of opposite sign then the roots of opposite sign

(iii) If a,b,c are of same sign then both the roots are negative

(iv) If a and c are of same sign differ from the sign of b then both the roots are positive

(v) If $a = c$ then the roots are reciprocal to each other

(vi) If $a + b + c = 0$ then the roots are $1, \dfrac{c}{a}$

(vii) If $a + c = b$ then the roots are $-1, -\dfrac{c}{a}$

Graphs of quadratic expression $ax^2 + bx + c$:

(i) If $a > 0$ and $b^2 - 4ac = 0$ then the graph of expression is upward parabola lies above the X-axis and touches the X-axis at one point

(ii) If $a > 0$ and $b^2 - 4ac > 0$ then the graph of expression is upward parabola intersects the X-axis at two points

(iii) If $a > 0$ and $b^2 - 4ac < 0$ then the graph of expression is upward parabola lies entirely above the X-axis

(iv) If $a < 0$ and $b^2 - 4ac = 0$ then the graph of expression is downward parabola lies below the X-axis and touches the X-axis at one point

(v) If $a < 0$ and $b^2 - 4ac > 0$ then the graph of expression is downward parabola intersects the X-axis at two points

(vi) If $a < 0$ and $b^2 - 4ac < 0$ then the graph of expression is downward parabola lies entirely below the X-axis

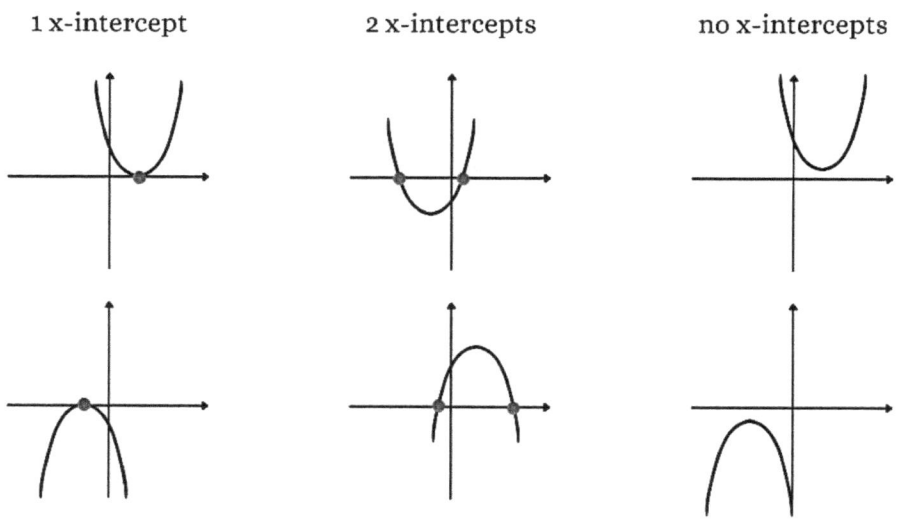

Maximum or Minimum Values of Quadratic Expressions:

(i) If $a > 0$ then the expression has minimum value at $x = \frac{-b}{2a}$ and the minimum value is given by $\frac{4ac-b^2}{4a}$

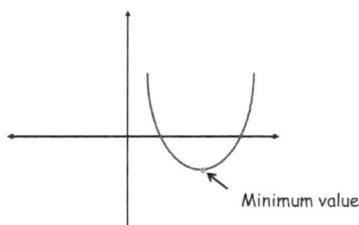

(ii) If $a < 0$ then the expression has maximum value at $x = \frac{-b}{2a}$ and the maximum value is given by $\frac{4ac-b^2}{4a}$

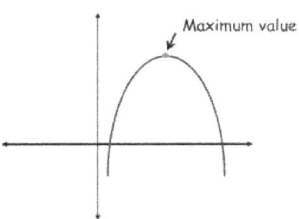

Signs of the Quadratic Expression:

Let α and β $(\alpha < \beta)$ be the real roots of $ax^2 + bx + c = 0$ then

(i) a and $ax^2 + bx + c$ have opposite sign when $\Delta > 0$ and $\alpha < x < \beta$

(ii) a and $ax^2 + bx + c$ have same sign when $\Delta > 0$ and $x < \alpha$ or $x > \beta$

Let $ax^2 + bx + c = 0$ has non real roots (i.e. $\Delta < 0$) then

(i) a and $ax^2 + bx + c$ have same sign $\forall x \in R$

(ii) a and $ax^2 + bx + c$ have same sign $\forall x \in \emptyset$

Locating the roots of Quadratic Equation under given condition:

Let Let α and β $(\alpha < \beta)$ be the real roots of $f(x) = ax^2 + bx + c = 0$ then

(i) Both the roots are less than some scalar k when $\Delta \geq 0, af(k) > 0$ and $\frac{-b}{2a} < k$

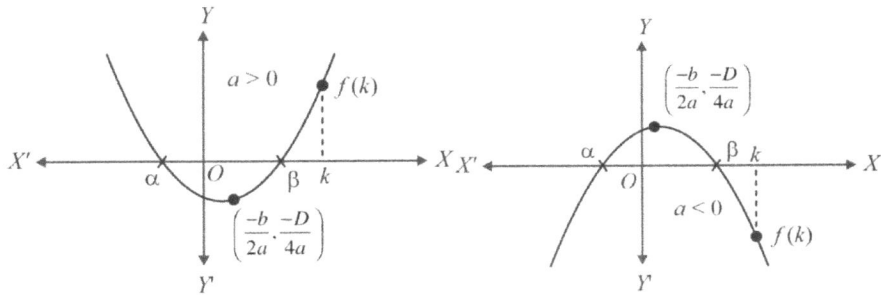

(ii) Both the roots are greater than some scalar k when $\Delta \geq 0, af(k) > 0$ and $\frac{-b}{2a} > k$

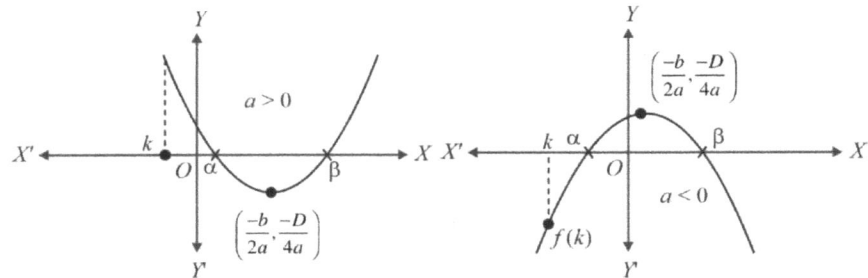

(iii) A scalar k lies between the roots when $\Delta > 0, af(k) < 0$

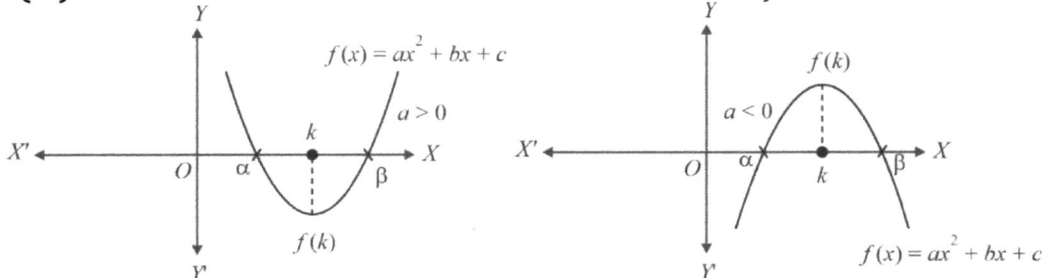

(iv) Both the roots lie between two scalars k_1 and k_2 ($k_1 < k_2$) when $\Delta \geq 0, af(k_1) > 0, af(k_2) > 0$ and $k_1 < \frac{-b}{2a} < k_2$

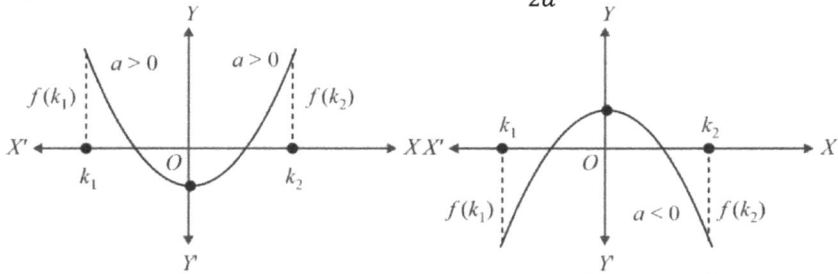

(v) Exactly one root is lie between k_1 and k_2 ($k_1 < k_2$) when $\Delta > 0, f(k_1).f(k_2) < 0$

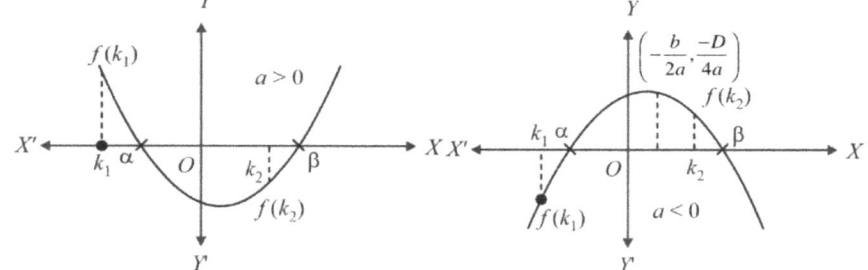

(vi) Two scalars k_1 and k_2 ($k_1 < k_2$) are lie between both the roots when $\Delta > 0, af(k_1) < 0, af(k_2) < 0$

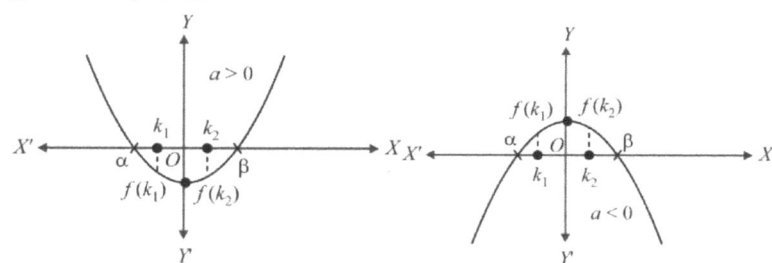

Transformed Equations of given Quadratic Equation:

Let $f(x) = ax^2 + bx + c = 0$ be a quadratic equation having roots α and β then the transformed equations of $f(x) = 0$ will be changed by changing the roots as follows

S.No.	New Roots	Transformed equation
1	$-\alpha, -\beta$	$f(-x) = 0$
2	$\dfrac{1}{\alpha}, \dfrac{1}{\beta}$	$f\left(\dfrac{1}{x}\right) = 0$
3	$k\alpha, k\beta$ where $k \neq 0$	$f\left(\dfrac{x}{k}\right) = 0$
4	$\dfrac{\alpha}{k}, \dfrac{\beta}{k}$ where $k \neq 0$	$f(kx) = 0$
5	$\alpha + k, \beta + k$	$f(x - k) = 0$
6	$\alpha - k, \beta - k$	$f(x + k) = 0$
7	α^2, β^2	$f(\sqrt{x}) = 0$
8	α^3, β^3	$f(\sqrt[3]{x}) = 0$

Common Root:

(i) The condition for one root is in common for both quadratic equations $a_1x^2 + b_1x + c_1 = 0$ and $a_2x^2 + b_2x + c_2 = 0$ is

$$(c_1a_2 - c_2a_1)^2 = (a_1b_2 - b_2a_1)(b_1c_2 - b_2c_1)$$

(ii) The condition for both the roots are in common for both quadratic equations $a_1x^2 + b_1x + c_1 = 0$ and $a_2x^2 + b_2x + c_2 = 0$ is $\dfrac{a_1}{a_2} = \dfrac{b_1}{b_2} = \dfrac{c_1}{c_2}$

14. THEORY OF EQUATIONS

Division Algorithm Theorem: If $f(x)$ and $g(x) (\neq 0)$ are two polynomials then there exists uniquely two polynomials $q(x)$ and $r(x)$ such that $f(x) = q(x).g(x) + r(x)$.

Here $q(x)$ is called as the quotient and $r(x)$ is called as the remainder of $f(x)$.

Remainder Theorem: If a polynomial $f(x)$ is divided by $x - a$ then the remainder is $f(a)$.

Factor Theorem: If a polynomial $f(x)$ is divisible by $x - a$ (i.e. $f(a) = 0$) then $x - a$ is a factor of $f(x)$.

Relation between the roots and the coefficients of Polynomial Equations:

(i) If α and β are the roots of quadratic equation $ax^2 + bx + c = 0$ then

$$S_1 \equiv \alpha + \beta = \frac{-b}{a} \; ; \; S_2 \equiv \alpha\beta = \frac{c}{a}$$

(ii) If α, β and γ are the roots of cubic equation $ax^3 + bx^2 + cx + d = 0$ then

$$S_1 \equiv \alpha + \beta + \gamma = \frac{-b}{a} \; ; \; S_2 \equiv \alpha\beta + \beta\gamma + \gamma\delta = \frac{c}{a} \; ; \; S_3 \equiv \alpha\beta\gamma = \frac{-d}{a}$$

(iii) If α, β, γ and δ are the roots of biquadratic equation $ax^4 + bx^3 + cx^2 + dx + e = 0$ then

$$S_1 \equiv \alpha + \beta + \gamma + \delta = \sum \alpha = \frac{-b}{a} \; ;$$

$$S_2 \equiv \alpha\beta + \beta\gamma + \gamma\delta + \alpha\gamma + \alpha\delta + \beta\delta = \sum \alpha\beta = \frac{c}{a} \; ;$$

$$S_3 \equiv \alpha\beta\gamma + \beta\gamma\delta + \alpha\beta\delta + \alpha\gamma\delta = \sum \alpha\beta\gamma = \frac{-d}{a} \; ; \; S_4 = \alpha\beta\gamma\delta = \frac{e}{a}$$

(iv) Continuing this process we have to say the relation about polynomial equation

If $\alpha_1, \alpha_2, \alpha_3, .., \alpha_n$ are the roots of polynomial equation $a_0 + a_1 x + a_2 x^2 + \cdots \ldots a_n x^n = 0$

Then

$$S_1 \equiv \sum \alpha_1 = \frac{-a_1}{a_0} \; ; \; S_2 \equiv \sum \alpha_1 \alpha_2 = \frac{a_2}{a_0} \; ; \; S_3 \equiv \sum \alpha_1 \alpha_2 \alpha_3 = \frac{-a_3}{a_0} \ldots \ldots$$

$$S_n \equiv \prod \alpha_1 = (-1)^n \frac{a_n}{a_0}$$

MATHEMATICS SUCCESS MANTRA

The Polynomial Equations, when the roots are given:

(i) The quadratic equation whose roots are α and β is $(x - \alpha)(x - \beta) = 0$

(ii) The cubic equation whose roots are α, β and γ is
$(x - \alpha)(x - \beta)(x - \gamma) = 0$

(iii) The biquadratic equation whose roots are α, β, γ and δ is
$(x - \alpha)(x - \beta)(x - \gamma)(x - \delta) = 0$

(iv) The polynomial equation whose roots are $\alpha_1, \alpha_2, \alpha_3, .., \alpha_n$ is
$(x - \alpha_1)(x - \alpha_2)(x - \alpha_3)......(x - \alpha_n) = 0$

Conditions for the equations when the roots are in progressions:

(i) The roots of a cubic equation are in A.P. then they can be taken as
$a - d, a, a + d$
The roots of a biquadratic equation are in A.P. then they can be taken as $a - 3d, a - d, a + d, a + 3d$

(ii) The roots of a cubic equation are in G.P. then they can be taken as
$\frac{a}{r}, a, ar$

The roots of a biquadratic equation are in A.P. then they can be taken as $\frac{a}{r^3}, \frac{a}{r}, ar, ar^3$

(iii) The roots of a cubic equation are in H.P. then they can be taken as
$\frac{1}{a - d}, \frac{1}{a}, \frac{1}{a + d}$
The roots of a biquadratic equation are in A.P. then they can be taken as $\frac{1}{a - 3d}, \frac{1}{a - d}, \frac{1}{a + d}, \frac{1}{a + 3d}$

Note: Let a polynomial equation has real coefficients.

(i) If $a + \sqrt{b}$ is a root of the equation then $a - \sqrt{b}$ is also root of the equation.

(ii) If $a + ib$ is a root of the equation then $a - ib$ is also root of the equation.

(iii) If $\sqrt{a} + \sqrt{b}$ is a root of the equation then $\sqrt{a} - \sqrt{b}, -\sqrt{a} + \sqrt{b}, -\sqrt{a} - \sqrt{b}$ are also the roots of the equation.

(iv) If $\sqrt{a} + i\sqrt{b}$ is a root of the equation then $\sqrt{a} - i\sqrt{b}, -\sqrt{a} + i\sqrt{b}, -\sqrt{a} - i\sqrt{b}$ are also the roots of the equation.

Multiple Roots: If a root α of a polynomial equation $f(x) = 0$ repeats k times then α is called the multiple root of order k. It is also of order k-1 of $f^1(x) = 0$

Newton's Method to find sum of the some powers of roots:

Let $\alpha, \beta, \gamma, \delta$ be the roots of byquadratic equation $x^4 + p_1x^3 + p_2x^2 + p_3x + p_4 = 0$.

Then the following formulae to find $S_1 = \alpha + \beta + \gamma + \delta$; $S_2 = \alpha^2 + \beta^2 + \gamma^2 + \delta^2$; $S_3 = \alpha^3 + \beta^3 + \gamma^3 + \delta^3$; $S_4 = \alpha^4 + \beta^4 + \gamma^4 + \delta^4$ ... ...are

$S_1 + p_1 = 0$; $S_2 + S_1 p_1 + 2p_2 = 0$; $S_3 + S_2 p_1 + S_1 p_2 + 3p_3 = 0$;

$S_4 + S_3 p_1 + S_2 p_2 + S_1 p_3 + 4p_4 = 0$ ... ...

Transformed Equations of given Polynomial Equation:

Let $f(x) = 0$ be a polynomial equation having real roots then the transformed equations of $f(x) = 0$ will be changed by changing the roots as follows

S.No.	New Roots	Transformed equation
1	Negatives of the roots of $f(x) = 0$	$f(-x) = 0$
2	Reciprocals of the roots of $f(x) = 0$	$f\left(\frac{1}{x}\right) = 0$
3	k times the roots of $f(x) = 0$ where $k \neq 0$	$f\left(\frac{x}{k}\right) = 0$
4	$\frac{1}{k}$ times the roots of $f(x) = 0$ where $k \neq 0$	$f(kx) = 0$
5	The roots of $f(x) = 0$ added by k	$f(x - k) = 0$
6	The roots of $f(x) = 0$ subtracted by k	$f(x + k) = 0$
7	Squares of the roots of $f(x) = 0$	$f(\sqrt{x}) = 0$
8	Cubes of the roots of $f(x) = 0$	$f(\sqrt[3]{x}) = 0$

Let $f(x) = 0$ be a polynomial equation of order n. To delete r^{th} term from beginning in the transformed equation of $f(x) = 0$, we have to diminish the roots of $f(x) = 0$ by h is given by $f^{n-r+1}(h) = 0$ i.e. $(n - r + 1)^{th}$ derivative of $f(h) = 0$.

Reciprocal Equation (R.E.): The polynomial equation $f(x) = 0$ is said to be reciprocal equation when it is unaltered by changing x by $\frac{1}{x}$.

Properties of R.E.: Let $f(x) = a_0 + a_1 x + a_2 x^2 + \cdots \ldots a_n x^n = 0$ be a R.E.

(i) $f(x) = 0$ said to be a R.E. of class one when $a_0 = a_n, a_1 = a_{n-1}, a_2 = a_{n-2}, \ldots..$

(ii) $f(x) = 0$ said to be R.E. of class two when $a_0 = -a_n, a_1 = -a_{n-1}, a_2 = -a_{n-2}$,

(iii) If $f(x) = 0$ an odd degree R.E. of class one then -1 is a root of the equation $f(x) = 0$

(iv) If $f(x) = 0$ an odd degree R.E. of class two then 1 is a root of the equation $f(x) = 0$

(v) If $f(x) = 0$ an even degree R.E. of class two then 1 and -1 both are the roots of the equation $f(x) = 0$

Descarte's Method to find the sign and nature of the roots:

Let $f(x) = 0$ be a polynomial equation of order n.

(i) If the number of sign changes of coefficients in $f(x) = 0$ is k then there are at most k positive roots.

(ii) If the number of sign changes of coefficients in $f(-x) = 0$ is k then there are at most k negitive roots.

(iii) If the number of sign changes of coefficients in both $f(x) = 0$ and $f(-x) = 0$ is k then there are at most k real roots and at least n-k imaginary roots.

15. PERMUTATIONS AND COMBINATIONS

PERMUTATIONS

Fundamental Principles

Multiplication Principle: A work done in m ways and another work done in n ways then the total work done by both works is mn ways.

Addition Principle: A work done in m ways and another work done in n ways then the total work done by either of the works is $m + n$ ways.

Factorial: The product of continuous first n natural numbers is defined as factorial n and denoted as $n!$ or $\llcorner n$

$$0! = 1; 1! = 1; 2! = 2; 3! = 6; 4! = 24; 5! = 120; 6! = 720; 7! = 5040 \ldots \ldots$$

Permutation: The number of arrangements of some or all things taken from given things.

Linear Permutation: If the objects are arranged in a line then the permutation is called as linear permutation.

Linear Permutations of given things:

(i) The number of linear permutations of r things taken from given n things is ${}^nP_r$

$${}^nP_r = \frac{n!}{(n-r)!}$$

(ii) The number of linear permutations of n things taken from given n things is ${}^nP_n$

$${}^nP_n = n!$$

Properties of ${}^nP_r$:

(i) ${}^nP_0 = 1$

(ii) ${}^nP_r = r \cdot {}^{n-1}P_{r-1}$

(iii) $\dfrac{{}^nP_r}{{}^nP_{r-1}} = n - r + 1$

(iv) $\dfrac{{}^nP_r}{{}^{n-1}P_{r-1}} = n$

(v) $\dfrac{n!}{(n-1)!} = n$ or $\dfrac{n!}{n} = (n-1)!$

MATHEMATICS SUCCESS MANTRA

Different types of linear permutations:

(i) The number of permutations of r things taken from given n things when a particular thing is always occur is $^{n-1}P_{r-1}$

(ii) The number of permutations of r things taken from given n things when a particular thing is always does not occur is $^{n-1}P_r$

(iii) The number of permutations of r things taken from given n things when k particular things are always occur is $^{n-k}P_{r-k}$

(iv) The number of permutations of r things taken from given n things when k particular things are always does not occur is $^{n-k}P_r$

(v) The number of permutations of r things taken from given n things when k particular things are always occur and m things are always does not occur is $^{n-k-m}P_{r-k}$

(vi) The number of permutations of given n things taken all at a time when m things always occur together is $m!\,(n - m + 1)!$

(vii) The number of permutations of given n things taken all at a time when m things always does not occur together is $n! - m!\,(n - m + 1)!$

(viii) The number of permutations of n different things when r things are in specified order is $^nP_{n-r}$ or $\dfrac{n!}{r!}$

Sum of the numbers:

(i) The sum of the $r - digited$ numbers by taking from n given things (excluding 0) is $^{n-1}P_{r-1} \times (Sum\ of\ the\ n\ digits) \times (1111\ldots r\ times)$

(ii) The sum of the $r - digited$ numbers by taking from n given things (including 0) is $^{n-1}P_{r-1} \times (Sum\ of\ the\ n\ digits) \times (1111\ldots r\ times) -$
$^{n-2}P_{r-2} \times (Sum\ of\ the\ n\ digits) \times (1111\ldots (r-1)\,times)$

(iii) The sum of the $n - digited$ numbers formed by n given things (excluding 0) is
$(n-1)! \times (Sum\ of\ the\ n\ digits) \times (1111\ldots n\ times)$

(iv) The sum of the $n - digited$ numbers formed by n given things (including 0) is
$(n-1)! \times (Sum\ of\ the\ n\ digits) \times (1111\ldots n\ times)$
$-(n-2)! \times (Sum\ of\ the\ n\ digits) \times (1111\ldots (n-1)\ times)$

Permutations when repetition is allowed:

(i) The number of permutations of r things taken from given n dissimilar things when repetition is allowed is n^r

(ii) The number of permutations of r things taken from given n dissimilar things when at least one thing is repeated is $n^r - {}^nP_r$

Properties:

(i) The number of permutations of n dissimilar things taken not more than r things at a time when repitition is allowed is

$$n + n^2 + n^3 + \cdots \ldots + n^r = \frac{n(n^r - 1)}{n - 1}$$

(ii) The number of permutations of n dissimilar things taken more than r things at a time when repitition is allowed is

$$n^{r+1} + n^{r+2} + n^{r+3} + \ldots \ldots + n^n = \frac{n(n^n - n^r)}{n - 1}$$

(iii) The number of permutations of n dissimilar things taken not more than r things at a time when repitition is not allowed is

${}^nP_0 + {}^nP_1 + {}^nP_2 + \ldots \ldots + {}^nP_r$

(iv) The number of permutations of n dissimilar things taken more than r things at a time when repitition is allowed is

${}^nP_{r+1} + {}^nP_{r+2} + {}^nP_{r+3} \ldots \ldots {}^nP_n$

Circular Permutation: If the objects are arranged in a circle or around a circular table then the permutation is called as circular permutation.

Properties:

(i) The number of circular permutations of n dissimilar things taken all at a time is $(n - 1)$

(ii) The number of permutations of n things into a garland or chains is $\frac{(n-1)!}{2}$

(iii) The number of circular permutations of r things taken from given n dissimilar things when both clockwise and anticlockwise directions considered is $\frac{{}^nP_r}{r}$

(iv) The number of circular permutations of r things taken from given n dissimilar things when any one of clockwise and anti clockwise directions considered is $\frac{{}^nP_r}{2r}$

Permutations of things in which some are alike, and the rest are different:

(i) The number of permutations of n things in which p things are alike and the rest are different is $\dfrac{n!}{p!}$

(ii) If p things are like of one kind, q things are alike of second kind, r things are alike of third kind then the number of permutations of all things is $\dfrac{(p+q+r)!}{p!q!r!}$

Rank of a word: If all the letters of a given word are permutated in all ways and arranged in alphabetical order then the numerical place of the given word is called as a rank of the word.

Dearrangements:

(i) The number of ways of placing the n letters in n addressed envelopes so that all letters go into correct envelop is 1

(ii) The number of ways of placing the n letters in n addressed envelopes so that one letter go into wrong envelop is 0

(iii) The number of ways of placing the n letters in n addressed envelopes so that all n letters go into wrong envelops is

$$n! \left[\dfrac{1}{2!} - \dfrac{1}{3!} + \dfrac{1}{4!} - \cdots \ldots + (-1)^n \dfrac{1}{n!}\right]$$

(iv) The number of ways of placing the n letters in n addressed envelopes so that all r letters go into wrong envelops is $^nP_r \left[\dfrac{1}{2!} - \dfrac{1}{3!} + \dfrac{1}{4!} - \cdots \ldots + (-1)^r \dfrac{1}{r!}\right]$

Palindromes: If there is no change in the word (Number) when the letters (Digits) are arranged in reverse way then the word is called palindrome.

Eg: MALAYALAM

Number of Palindromes:

(i) The number of palindromes r distinct letters (digits) taken from given n distinct letters (digits excluding zero) = $\begin{cases} n^{r/2} \text{ when } r \text{ is even} \\ n^{\frac{r+1}{2}} \text{ when } r \text{ is odd} \end{cases}$

(ii) The number of palindromes r distinct letters (digits) taken from given n distinct letters (digits including zero) = $\begin{cases} (n-1)n^{\frac{r-2}{2}} \text{ when } r \text{ is even} \\ (n-1)n^{\frac{r-1}{2}} \text{ when } r \text{ is odd} \end{cases}$

COMBINATIONS

Combination: The number of ways of selecting some or all things from given things is called as combination.

Combination of given things:

(i) The number of ways of selecting r things from given n things is $^nC_r = \dfrac{n!}{r!(n-r)!}$

(ii) The number of ways of selecting of all n given things $^nC_n = 1$

Properties of : ^nC_r

(i) $^nC_r = {^nC_{n-r}}$

(ii) $^nC_r + {^nC_{r-1}} = {^{n+1}C_r}$

(iii) $\dfrac{^nC_r}{^nC_{r-1}} = \dfrac{n-r+1}{r}$

(iv) If $^nC_r = {^nC_s}$ then $r = s$

(v) $^nC_0 + {^nC_1} + {^nC_2} + {^nC_3} + \cdots \ldots + {^nC_n} = 2^n$

(vi) $^nC_0 - {^nC_1} + {^nC_2} - {^nC_3} + \cdots \ldots + (-1)^n \, {^nC_n} = 0$

(vii) $^nC_0 + {^nC_2} + {^nC_4} + \cdots \ldots = {^nC_1} + {^nC_3} + {^nC_5} + \cdots \ldots = 2^{n-1}$

(viii) $a.\,{^nC_0} + (a+d)\,{^nC_1} + (a+2d)\,{^nC_2} + \cdots \ldots + (a+nd)\,{^nC_n} = (2a+nd)2^{n-1}$

(ix) $\dfrac{^nP_r}{^nC_r} = r!$

(x) If ^nC_r is greatest value then $r = \begin{cases} \dfrac{n}{2} & \text{when } n \text{ is even} \\ \dfrac{n-1}{2} \text{ or } \dfrac{n+1}{2} & \text{when } n \text{ is odd} \end{cases}$

(xi) **Vandermonde's Theorem:**

$$^mC_0 \cdot {^nC_r} + {^mC_1} \cdot {^nC_{r-1}} + {^mC_2} \cdot {^nC_{r-2}} + \cdots \ldots + {^mC_r} \cdot {^nC_0} = {^{m+n}C_r}$$

Different types of Combinations:

(i) The number of combinations of r things taken from given n things when a particular thing is always occur is $^{n-1}C_{r-1}$

(ii) The number of combinations of r things taken from given n things when a particular thing is always does not occur is $^{n-1}C_r$

(iii) The number of combinations of r things taken from given n things when k particular things are always occur is $^{n-k}C_{r-k}$

(iv) The number of combinations of r things taken from given n things when k particular things are always does not occur is $^{n-k}C_r$

(v) The number of combinations of r things taken from given n things when k particular things are always occur and m things are always does not occur is $^{n-k-m}C_{r-k}$

(vi) The number of combinations of r things taken from given n things when p particular things are not together is $^nC_r - {^{n-p}C_{r-p}}$

(vii) The number of combinations of r consecutive things from n given things in a row is $n - r + 1$

(viii) The number of combinations of r consecutive things from n given things around a circle is $\begin{cases} n \text{ when } r < n \\ 1 \text{ when } r = n \end{cases}$

Distribution of things:

(i) The number of ways of dividing $m + n$ things into two equal groups containing m and n things each is $^{m+n}C_m = {^{m+n}C_n} = \frac{(m+n)!}{m!n!}$

(ii) The number of ways of distributing $m + n$ things to two persons such that one person gets m things and another person gets n things each is $\frac{(m+n)!}{m!n!} \times 2!$

(iii) The number of ways of dividing mn things into m groups, each group containing n things is $\frac{(mn)!}{(n!)^m} \times \frac{1}{m!}$

(iv) The number of ways of distributing mn things to m persons such that each person gets n things is $\frac{(mn)!}{(n!)^m}$

Combinations in Geometry:

(i) The number of straight lines can be formed by n points in a plane is $^nC_2$

(ii) The number of triangles can be formed by n points in a plane is $^nC_3$

(iii) The number of quadrilaterals can be formed by n points in a plane is $^nC_4$

(iv) The number of polygons with r vertices can be formed by n points in a plane is $^nC_5$

(v) The number of diagonals of $n-sided$ polygon is $\frac{n(n-3)}{2}$ or $^nC_2-n$

(vi) The number of parallelograms can be formed by intersecting of a set of m parallel lines and a set of n parallel lines is $^mC_2 \times {}^nC_2$

(v) If k points are collinear out of n points in a plane then the number of straight lines that can be formed joining them is $^nC_2 - {}^kC_2 + 1$

(vi) If k points are collinear out of n points in a plane then the number of triangles that can be formed joining them is $^nC_3 - {}^kC_3$

(vii) The number of intersecting points of n straight lines in a plane is $^nC_2$

(viii) The number of intersecting points of n circles in a plane is $^nC_2 \times 2$

(ix) The number of intersecting points of n straight lines and m circles in a plane is $^nC_2 \times {}^mC_2 \times 2$

(x) The number of intersecting points of diagonals of $n-sided$ polygon in which the diagonals lie completely inside the polygon is $^nC_4$

(xi) The number of straight lines joining the points of intersections of n straight lines in which no two are parallel and no three are concurrent is $\frac{n(n-1)(n-2)(n-3)}{8}$ where $n > 3$

(xii) The number of triangles whose angular points are the angular points of $n-sided$ polygon in which none of whose sides are the sides of the polygon is $\frac{n(n-4)(n-5)}{6}$

Combinations in chess board:

(i) The number of rectangles of any size in a square of size $n \times n$ is $\sum_{r=1}^{n}(r^3)$

(ii) The number of squares of any size in a square of size $n \times n$ is $\sum_{r=1}^{n}(r^2)$

(iii) The number of rectangles of any size in a rectangle of size $n \times p$ $(n < p)$ is $\frac{np(n+1)(p+1)}{4}$

(iv) The number of rectangles of any size in a rectangle of size $n \times p$ $(n < p)$ is $\sum_{r=1}^{n}(n+1-r)(p+1-r)$

(v) The number of rectangles on a chessboard (including squares) is $\sum_{r=1}^{8}(r^3) = 1296$ or $^9C_2 \times {}^9C_2 = 1296$

(vi) The number of squares on a chessboard is 204

(vii) The number of rectangles which are not squares on a chessboard is 1092

MATHEMATICS SUCCESS MANTRA

Combinations of alike things and distinct things:

(i) The number of selections of none or more things from $p_1 + p_2 + p_3 + \ldots + p_n$ things in which p_1 things are alike of one kind, p_2 things are alike of second kind, p_3 things are alike of third kind and so on p_k things are alike of kth kind is $(p_1 + 1) + (p_2 + 1) + (p_3 + 1) + \cdots \ldots + (p_n + 1)$

(ii) The number of selections of one or more things from $p_1 + p_2 + p_3 + \ldots + p_n$ things in which p_1 things are alike of one kind, p_2 things are alike of second kind, p_3 things are alike of third kind and so on p_k things are alike of kth kind is $(p_1 + 1) + (p_2 + 1) + (p_3 + 1) + \ldots + (p_n + 1) - 1$

(iii) The number of selections of none or more things from n distinct things is $^nC_0 + {}^nC_1 + {}^nC_2 + {}^nC_3 + \ldots + {}^nC_n = 2^n$

(iv) The number of selections of none or more things from n distinct things is $^nC_1 + {}^nC_2 + {}^nC_3 + \ldots + {}^nC_n = 2^n - 1$

(v) The number of selections of none or more things from $p_1 + p_2 + p_3 + \ldots + p_n + q$ things in which p_1 things are alike of one kind, p_2 things are alike of second kind, p_3 things are alike of third kind and so on p_k things are alike of kth kind and the remaining q things are distinct is

$(p_1 + 1) + (p_2 + 1) + (p_3 + 1) + \cdots \ldots + (p_n + 1)2^q$

(vi) The number of selections of one or more things from $p_1 + p_2 + p_3 + \cdots \ldots + p_n + q$ things in which p_1 things are alike of one kind, p_2 things are alike of second kind, p_3 things are alike of third kind and so on p_k things are alike of kth kind and the remaining q things are distinct is

$(p_1 + 1) + (p_2 + 1) + (p_3 + 1) + \cdots \ldots + (p_n + 1)2^q - 1$

(vii) The number of ways of dividing n identical things to r persons such that each one can receive 0 or 1 or 2 or more items is $^{n+r-1}C_{r-1}$

(viii) The number of ways of dividing n identical things to r persons such that each one can receive at least one items is $^{n-1}C_{r-1}$

Divisors and sum of the divisors of a number:

(i) The total number of divisors of a number $p_1^{\alpha_1} \cdot p_2^{\alpha_2} \cdot p_3^{\alpha_3} \ldots \ldots p_n^{\alpha_n}$ where $p_1, p_2, p_3, \ldots \ldots, p_n$ are primes and $\alpha_1, \alpha_2, \alpha_3, \ldots \ldots, \alpha_n$ are positive integers is $(\alpha_1 + 1)(\alpha_2 + 1)(\alpha_3 + 1) \ldots \ldots (\alpha_n + 1)$

(ii) The total number of proper divisors (1 and itself the number) of a number $p_1^{\alpha_1} \cdot p_2^{\alpha_2} \cdot p_3^{\alpha_3} \ldots \ldots p_n^{\alpha_n}$ where $p_1, p_2, p_3, \ldots \ldots, p_n$ are primes and $\alpha_1, \alpha_2, \alpha_3, \ldots \ldots, \alpha_n$ are positive integers is $(\alpha_1 + 1)(\alpha_2 + 1)(\alpha_3 + 1) \ldots \ldots (\alpha_n + 1) - 2$

(iii) The sum of all divisors of a number $p_1^{\alpha_1} \cdot p_2^{\alpha_2} \cdot p_3^{\alpha_3} \ldots \ldots p_n^{\alpha_n}$ where

$p_1, p_2, p_3, \ldots, p_n$ are primes and $\alpha_1, \alpha_2, \alpha_3, \ldots, \alpha_n$ are positive integers is $\left(\frac{p_1^{\alpha_1+1}-1}{p_1-1}\right)\left(\frac{p_2^{\alpha_2+1}-1}{p_2-1}\right)\left(\frac{p_3^{\alpha_3+1}-1}{p_3-1}\right)\ldots\left(\frac{p_n^{\alpha_n+1}-1}{p_n-1}\right)$

(iv) The total number of odd divisors of a number $2^{\alpha_1} \cdot p_2^{\alpha_2} \cdot p_3^{\alpha_3} \ldots p_n^{\alpha_n}$ where $p_2, p_3, \ldots, p_n$ are odd primes and $\alpha_1, \alpha_2, \alpha_3, \ldots, \alpha_n$ are positive integers is $(\alpha_2 + 1)(\alpha_3 + 1)\ldots(\alpha_n + 1)$

(v) The total number of even divisors of a number $2^{\alpha_1} \cdot p_2^{\alpha_2} \cdot p_3^{\alpha_3} \ldots p_n^{\alpha_n}$ where $p_2, p_3, \ldots, p_n$ are odd primes and $\alpha_1, \alpha_2, \alpha_3, \ldots, \alpha_n$ are positive integers is $\alpha_1(\alpha_2 + 1)(\alpha_3 + 1)\ldots(\alpha_n + 1)$

(vi) The number of ways that the number $p_1^{\alpha_1} \cdot p_2^{\alpha_2} \cdot p_3^{\alpha_3} \ldots p_n^{\alpha_n}$ where $p_1, p_2, p_3, \ldots, p_n$ are primes and $\alpha_1, \alpha_2, \alpha_3, \ldots, \alpha_n$ are positive integers can be resolve into two factors is

$$\begin{cases} \frac{(\alpha_1+1)(\alpha_2+1)(\alpha_3+1)\ldots(\alpha_n+1)}{2} & \text{where the number is not a perfect square} \\ \frac{(\alpha_1+1)(\alpha_2+1)(\alpha_3+1)\ldots(\alpha_n+1)+1}{2} & \text{where the number is a perfect square} \end{cases}$$

Integral Solutions:

(i) The number of positive integral solutions of the equation $x_1 + x_2 + x_3 \ldots + x_r = n$ is $^{n-1}C_{r-1}$

(ii) The number of non negative integral solutions of the equation $x_1 + x_2 + x_3 + \cdots + x_r = n$ is $^{n+r-1}C_{r-1}$

(iii) The number of non negative integral solutions of $x_1 + x_2 + x_3 + \cdots + x_r \leq n$ is $^{n+r}C_r$

Exponent of P (prime) in $n!$ $(n \in N)$:

(i) The exponent (highest power) of P in $n!$ is $\left[\frac{n}{P}\right] + \left[\frac{n}{P^2}\right] + \left[\frac{n}{P^3}\right] + \ldots$ where [.] is a greatest integer function

(ii) The number of zeroes at the end of the number $n! = 2^\alpha 3^\beta 5^\gamma 7^\delta \ldots$ is γ

16. BINOMIAL THEOREM

Binomial Expansions for the index of positive integers :

Let x, a be real numbers and n be a positive integer then

$$(x+a)^n = {}^nC_0 x^n + {}^nC_1 x^{n-1}a + {}^nC_2 x^{n-2}a^2 + {}^nC_3 x^{n-3}a^3 + \ldots + {}^nC_n a^n = \sum_{r=0}^{n} {}^nC_r x^{n-r} a^r$$

Here ${}^nC_0, {}^nC_1, {}^nC_2, {}^nC_3, \ldots\ldots\ldots, {}^nC_n$ are called as binomial coefficients.

They simply represented by $C_0, C_1, C_2, C_3, \ldots\ldots, C_n$

Note:

(i) $(x-a)^n = {}^nC_0 x^n - {}^nC_1 x^{n-1}a + {}^nC_2 x^{n-2}a^2 - {}^nC_3 x^{n-3}a^3 + \ldots + (-1)^n \, {}^nC_n a^n$

$\qquad = \sum_{r=0}^{n}(-1)^r \, {}^nC_r x^{n-r} a^r$

(ii) $(1+x)^n = {}^nC_0 + {}^nC_1 x + {}^nC_2 x^2 + {}^nC_3 x^3 + \ldots\ldots\ldots + {}^nC_n x^n = \sum_{r=0}^{n} {}^nC_r x^r$

(iii) $(1-x)^n = {}^nC_0 - {}^nC_1 x + {}^nC_2 x^2 - {}^nC_3 x^3 + \ldots\ldots\ldots + (-1)^n \, {}^nC_n x^n = \sum_{r=0}^{n}(-1)^r \, {}^nC_r x^r$

Number of terms:

(i) The number of terms in the expansion of

$(x+a)^n = {}^nC_0 x^n + {}^nC_1 x^{n-1}a + {}^nC_2 x^{n-2}a^2 + {}^nC_3 x^{n-3}a^3 + \ldots + {}^nC_n a^n = \sum_{r=0}^{n} {}^nC_r x^{n-r} a^r$ is n+1

(ii) The number of terms in the expansion of trinomial $(a+b+c)^n$ is $\frac{(n+1)(n+2)}{2}$

(iii) The number of terms in the expansion of multinomial $(a_1 + a_2 + \cdots \ldots a_r)^n$ is ${}^{n+r-1}C_{r-1}$

General term: The $(r+1)^{th}$ term in a binomial expansion is considered as general term

(i) In the expansion of $(x+a)^n$, the general term is $T_{r+1} = \sum_{r=0}^{n} {}^nC_r x^{n-r} a^r$

(ii) In the expansion of $(x-a)^n$, the general term is $T_{r+1} = \sum_{r=0}^{n}(-1)^r \, {}^nC_r x^{n-r} a^r$

Coefficient of x^r :

To find the coefficient of x^k in the expansion of $\left(ax^p + \frac{b}{x^q}\right)^n$, Take $r = \frac{np-k}{p+q}$

Term independent of x :

To find the term independent of x in the expansion of $\left(ax^p + \frac{b}{x^q}\right)^n$,

Take $r = \frac{np}{p+q}$

Middle Term(s):

(i) If n is even then there is only one middle term in the expansion of $(x+a)^n$.

Middle term $= T_{\frac{n}{2}+1}$

(ii) If n is odd then there are two middle terms in the expansion of $(x+a)^n$.

Middle terms $= T_{\frac{n+1}{2}}$ and $T_{\frac{n+3}{2}}$

Largest Binomial coefficient(s):

(i) If n is even, then there is only one largest binomial coefficient in the expansion of $(x+a)^n$

Largest binomial coefficient $= {}^nC_{\frac{n}{2}}$

(ii) If n is odd then there are two largest binomial coefficients in the expansion of $(x+a)^n$.

Largest binomial coefficients $= {}^nC_{\frac{n-1}{2}}$ and ${}^nC_{\frac{n+1}{2}}$

Numerically Greatest Term(s):

To find the numerically greatest term in $(a+x)^n$, first of all convert the expansion in the form of $a^n(1+\frac{x}{a})^n$ and take $X = \frac{x}{a}$

Now, Take $r = \left[\frac{(n+1)|X|}{1+|X|}\right]$

(i) If $\frac{(n+1)|X|}{1+|X|}$ is not an integer then there is only one numerically greatest term in the expansion. That is T_{r+1}

(ii) If $\frac{(n+1)|X|}{1+|X|}$ is an integer then there are two numerically greatest terms in the expansion. They are T_r and T_{r+1} such that $|T_r| = |T_{r+1}|$

Properties of Binomial coefficients :

(i) $C_r = C_{n-r}$

(ii) $^nC_r + {}^nC_{r-1} = {}^{n+1}C_r$

(iii) $\dfrac{C_r}{C_{r-1}} = \dfrac{n-r+1}{r}$

(iv) If $^nC_r = {}^nC_s$ then $r = s$

(v) $C_0 + C_1 + C_2 + C_3 + \cdots \ldots + C_n = 2^n$

(vi) $C_0 - C_1 + C_2 - C_3 + \cdots \ldots + (-1)^n C_n = 2^n$

(vii) $C_0 + C_2 + C_4 + \cdots \ldots = C_1 + C_3 + C_5 + + \cdots \ldots = 2^n$

(viii) $C_0 C_r + C_1 C_{r+1} + C_2 C_{r+2} + \cdots \ldots + C_{n-r} C_n = {}^{2n}C_{n-r} = {}^{2n}C_{n-r} = \dfrac{(2n)!}{(n-r)!(n+r)!}$

(ix) $C_0^2 + C_1^2 + C_2^2 + C_3^2 + \cdots \ldots + C_n^2 = {}^{2n}C_n$

(x) $C_0^2 - C_1^2 + C_2^2 - C_3^2 + \cdots \ldots + (-1) \; C_n^2 = \begin{cases} (-1)^{n/2} \, {}^nC_{n/2} \text{ when } n \text{ is even} \\ 0 \text{ when } n \text{ is odd} \end{cases}$

(xi) $a.C_0 + (a+d).C_1 + (a+2d).C_2 + \cdots \ldots + (a+nd).C_n = (2a+nd)2^{n-1}$

(xii) $a.C_0 - (a+d).C_1 + (a+2d).C_2 - \cdots \ldots + (-1)^n (a+nd).C_n = 0$

(xiii) $a.C_0^2 + (a+d).C_1^2 + (a+2d).C_2^2 + \cdots \ldots + (a+nd).C_n^2 = \left(\dfrac{2a+nd}{2}\right){}^{2n}C_n$

(xiv) $\sum_{r=0}^{n} r. \, {}^nC_r = n.2^{n-1}$

(xv) $\sum_{r=0}^{n} r(r-1) \, {}^nC_r = n(n-1)2^{n-2}$

(xvi) $\sum_{r=0}^{n} r(r-1)(r-2) \, {}^nC_r = n(n-1)(n-2)2^{n-3}$

(xvii) $C_0 + \dfrac{C_1}{2}x + \dfrac{C_2}{3}x^2 + \cdots \ldots + \dfrac{C_n}{n}x^n = \dfrac{(1+x)^{n+1} - 1}{(n+1)x}$

(xviii) $(C_0 + C_1)(C_1 + C_2) \ldots \ldots (C_{n-1} + C_n) = \dfrac{(C_0.C_1.C_2 \ldots .C_n)}{n!}(n+1)^n$

(xix) If ^nC_r is greatest value then $r = \begin{cases} \dfrac{n}{2} \text{ when } n \text{ is even} \\ \dfrac{n-1}{2} \text{ or } \dfrac{n+1}{2} \text{ when } n \text{ is odd} \end{cases}$

(xx) **Vandermonde's Theorem:**

$^mC_0. \, {}^nC_r + {}^mC_1. \, {}^nC_{r-1} + {}^mC_2. \, {}^nC_{r-2} + \cdots \ldots + {}^mC_r. \, {}^nC_0 = {}^{m+n}C_r$

Sum of the coefficients : Let $f(x)$ be a binomial expansion in x then

    (i) The sum of the coefficients $= f(1)$

    (ii) The sum of the coefficients of even powers of x (odd terms) $= \dfrac{f(1)+f(-1)}{2}$

    (iii) The sum of the coefficients of odd powers of x (even terms) $= \dfrac{f(1)-f(-1)}{2}$

Binomial Expansions for the index of negative integers and rational numbers:

(i) $(1-x)^{-n} = 1 + \dfrac{n}{1!}x + \dfrac{n(n+1)}{2!}x^2 + \dfrac{n(n+1)(n+2)}{3!}x^3 + \cdots \ldots \ldots \infty$ where $|x| < 1$

    General term, $T_{r+1} = \dfrac{n(n+1)(n+2)\ldots[n+(r-1)]}{r!} x^r$

(ii) $(1+x)^{-n} = 1 - \dfrac{n}{1!}x + \dfrac{n(n+1)}{2!}x^2 - \dfrac{n(n+1)(n+2)}{3!}x^3 + \cdots \ldots \ldots \infty$ where $|x| < 1$

    General term, $T_{r+1} = (-1)^r \dfrac{n(n+1)(n+2)\ldots[n+(r-1)]}{r!} x^r$

(iii) $(1-x)^{-p/q} = 1 + \dfrac{p}{1!}\left(\dfrac{x}{q}\right) + \dfrac{p(p+q)}{2!}\left(\dfrac{x}{q}\right)^2 + \dfrac{p(p+q)(p+2q)}{3!}\left(\dfrac{x}{q}\right)^3 + \cdots \ldots \ldots \infty$

    where $|x| < 1$

    General term, $T_{r+1} = \dfrac{p(p+q)(p+2q)\ldots[p+(r-1)q]}{r!} \left(\dfrac{x}{q}\right)^r$

(iv) $(1+x)^{-p/q} = 1 - \dfrac{p}{1!}\left(\dfrac{x}{q}\right) + \dfrac{p(p+q)}{2!}\left(\dfrac{x}{q}\right)^2 - \dfrac{p(p+q)(p+2q)}{3!}\left(\dfrac{x}{q}\right)^3 + \cdots \ldots \ldots \infty$

    where $|x| < 1$

    General term, $T_{r+1} = (-1)^r \dfrac{p(p+q)(p+2q)\ldots[p+(r-1)q]}{r!} \left(\dfrac{x}{q}\right)^r$

(v) $(1+x)^{p/q} = 1 + \dfrac{p}{1!}\left(\dfrac{x}{q}\right) + \dfrac{p(p-q)}{2!}\left(\dfrac{x}{q}\right)^2 + \dfrac{p(p-q)(p-2q)}{3!}\left(\dfrac{x}{q}\right)^3 + \cdots \ldots \ldots \infty$

    where $|x| < 1$

    General term, $T_{r+1} = \dfrac{p(p-q)(p-2q)\ldots[p-(r-1)q]}{r!} \left(\dfrac{x}{q}\right)^r$

(vi) $(1-x)^{p/q} = 1 - \dfrac{p}{1!}\left(\dfrac{x}{q}\right) + \dfrac{p(p-q)}{2!}\left(\dfrac{x}{q}\right)^2 - \dfrac{p(p-q)(p-2q)}{3!}\left(\dfrac{x}{q}\right)^3 + \cdots \ldots \ldots \infty$

    where $|x| < 1$

    General term, $T_{r+1} = (-1)^r \dfrac{p(p-q)(p-2q)\ldots[p-(r-1)q]}{r!} \left(\dfrac{x}{q}\right)^r$

Note: (i) $(1-x)^{-1} = 1 + x + x^2 + x^3 + \cdots \ldots \ldots \infty$

    (ii) $(1+x)^{-1} = 1 - x + x^2 - x^3 + \cdots \ldots \ldots \infty$

    (iii) $(1-x)^{-2} = 1 + 2x + 3x^2 + 4x^3 + \cdots \ldots \ldots \infty$

    (iv) $(1+x)^{-2} = 1 - 2x + 3x^2 - 4x^3 + \cdots \ldots \ldots \infty$

    (v) $(1-x)^{-3} = 1 + 3x + 6x^2 + 10x^3 + \cdots \ldots \ldots \infty$

    (vi) $(1+x)^{-3} = 1 - 3x + 6x^2 - 10x^3 + \cdots \ldots \ldots \infty$

TRIGONOMETRY

17. TRIGONOMETRIC RATIOS AND IDENTITIES

Measure of an angle: Angle is the amount of rotation of a given ray from the initial side to its terminal side about vertex.

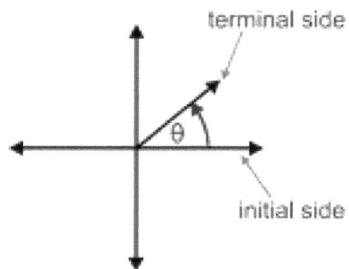
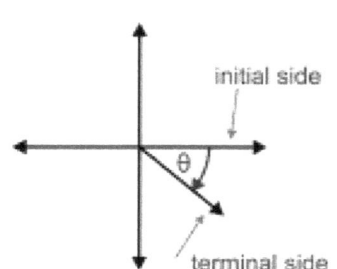

Degree Measure: The angle is said to be a measure of 1 degree, if a rotation from the initial side to terminal side is $\frac{1}{360}^{th}$ of a revolution. It is represented by 1^0.

$$1^0 (1\ degree) = 60' (60\ minutes)\ ;\ 1' (1\ minute) = 60'' (60\ seconds)$$

Radian Measure: The angle is said to be a measure of 1 radian (1^c), if it is the angle measures at the center by an arc of length 1 unit in a circle.

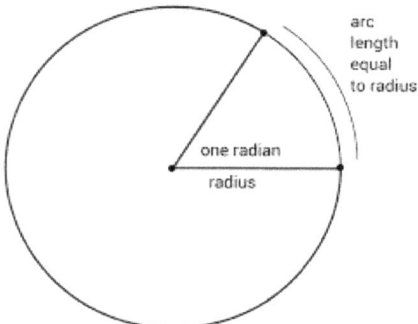

Relation between degree and radian measures:

$$Radian\ Measure = \frac{\pi}{180} \times Degree\ Measure$$

$$Degree\ Measure = \frac{180}{\pi} \times Radian\ Measure$$

$$1^0 = \frac{\pi}{180}\ Radians = 0.01746\ (approx)$$

$$1^c = \frac{180}{\pi}\ Degrees = 57^0 16'\ (approx)$$

Pythagorean Theorem:

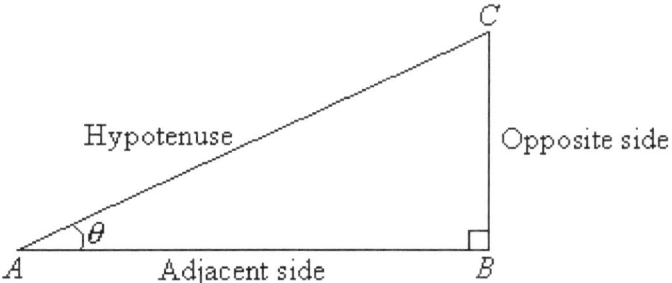

Let ABC be a right angled triangle with right angle at B. Then the Pythagorean theorem stated as

$AC^2 = AB^2 + BC^2$ i.e. $Hypotenuse^2 = Side^2 + Side^2$

Pythagorean Triplets:

(3,4,5),(5,12,13),(7,24,25),(8,15,17),(9,40,41),(11,60,61),(16,63,65) ....

In all triplets, the largest side is the hypotenuse.

Trigonometric Ratios for acute angle:

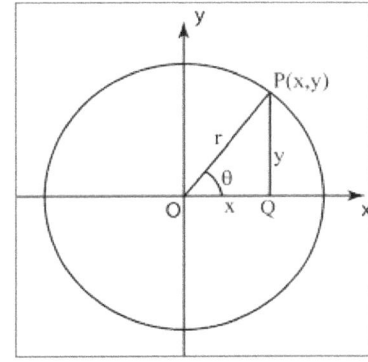

Let XOY be a rectangular cartesian co-ordinate system.

Let $P(x, y)$ be a point in quadrant 1 with $OP = r$.

Draw a circle with centre O and radius r.

Draw a perpendicular PQ to X-axis. Then $OQ = x$ and $PQ = y$.

Let OP makes an angle θ with the positive X-axis in anticlockwise direction.

Then the trigonometric ratios are defined as follows

$$Sin\ \theta = \frac{Opposite\ side\ to\ \theta}{Hypotenuse}$$

$$Cos\ \theta = \frac{Adjacent\ side\ to\ \theta}{Hypotenuse}$$

$$Tan\ \theta = \frac{Opposite\ side\ to\ \theta}{Adjacent\ side\ to\ \theta}$$

$$Cot\ \theta = \frac{Adjacent\ side\ to\ \theta}{Oppposite\ side\ to\ \theta}$$

$$Sec\ \theta = \frac{Hypotenuse}{Adjacent\ side\ to\ \theta}$$

$$Cosec\ \theta = \frac{Hypotenuse}{Opposite\ side\ to\ \theta}$$

Signs of the trigonometric ratios in various quadrants:

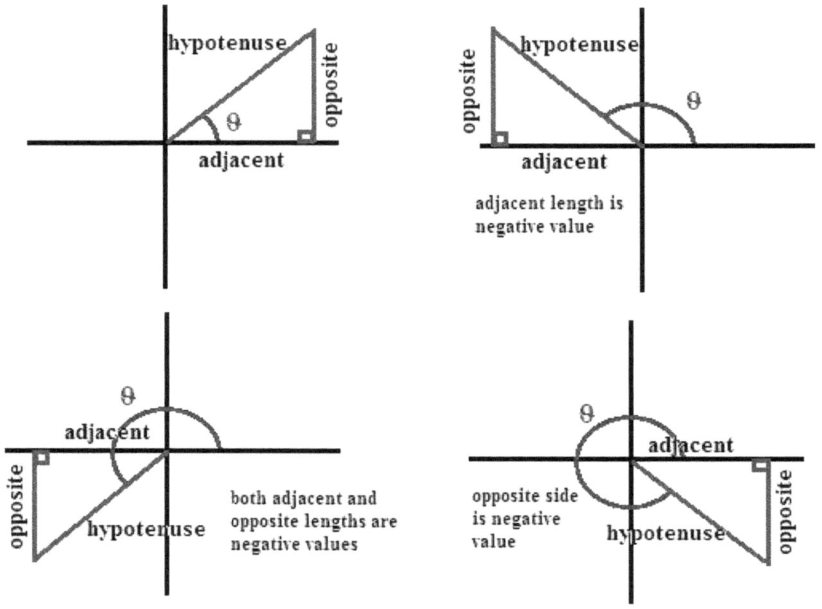

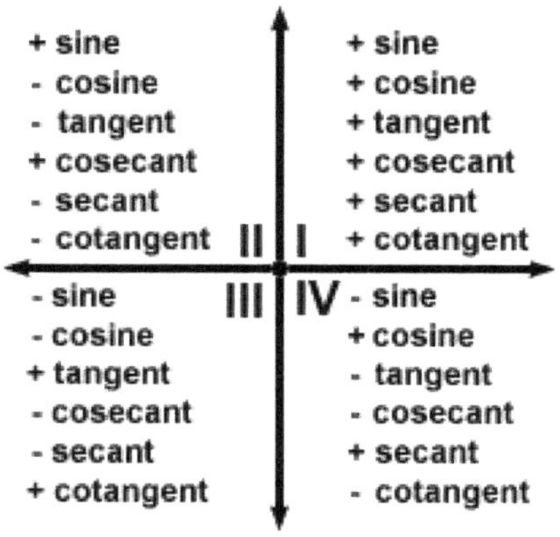

Quadrant	$\sin\alpha$	$\cos\alpha$	$\tan\alpha$	$\cot\alpha$	$\sec\alpha$	$\csc\alpha$
I	+	+	+	+	+	+
II	+	−	−	−	−	+
III	−	−	+	+	−	−
IV	−	+	−	−	+	−

The values of trigonometric ratios in various quadrants:

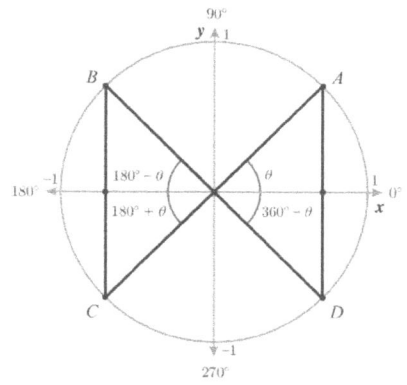

	$90°-\theta$	$90°+\theta$	$180°-\theta$	$180°+\theta$	$270°-\theta$	$270°+\theta$	$360°-\theta$	$360°+\theta$
Sin	$Cos\,\theta$	$Cos\,\theta$	$Sin\,\theta$	$-Sin\,\theta$	$-Cos\,\theta$	$-Cos\,\theta$	$-Sin\,\theta$	$Sin\,\theta$
Cos	$Sin\,\theta$	$-Sin\,\theta$	$-Cos\,\theta$	$-Cos\,\theta$	$-Sin\,\theta$	$Sin\,\theta$	$Cos\,\theta$	$Cos\,\theta$
Tan	$Cot\,\theta$	$-Cot\,\theta$	$-Tan\,\theta$	$Tan\,\theta$	$Cot\,\theta$	$-Cot\,\theta$	$-Tan\,\theta$	$Tan\,\theta$
Cot	$Tan\,\theta$	$-Tan\,\theta$	$-Cot\,\theta$	$Cot\,\theta$	$Tan\,\theta$	$-Tan\,\theta$	$-Cot\,\theta$	$Cot\,\theta$
Sec	$Cosec\,\theta$	$-Cosec\,\theta$	$-Sec\,\theta$	$-Sec\,\theta$	$-Cosec\,\theta$	$Cosec\,\theta$	$Sec\,\theta$	$Sec\,\theta$
Cosec	$Sec\,\theta$	$Sec\,\theta$	$Cosec\,\theta$	$-Cosec\,\theta$	$-Sec\,\theta$	$-Sec\,\theta$	$-Cosec\,\theta$	$Cosec\,\theta$

The values of trigonometric ratios for negative angle:

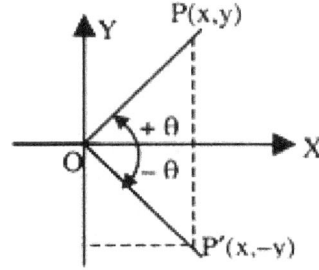

$-\theta$ is in fourth quadrant.

$Sin(-\theta) = -Sin\,\theta;\ Cos(-\theta) = Cos\,\theta;\ Tan(-\theta) = -Tan\,\theta$

$Cot(-\theta) = -Cot\,\theta;\ Sec(-\theta) = Sec\,\theta;\ Cosec(-\theta) = -Cosec\,\theta$

Trigonometric ratios for different values:

	$0°$	$30°$	$45°$	$60°$	$90°$	$180°$	$270°$	$360°$
$Sin\,\theta$	0	$\dfrac{1}{2}$	$\dfrac{1}{\sqrt{2}}$	$\dfrac{\sqrt{3}}{2}$	1	0	-1	0
$Cos\,\theta$	1	$\dfrac{\sqrt{3}}{2}$	$\dfrac{1}{\sqrt{2}}$	$\dfrac{1}{2}$	0	-1	0	1
$Tan\,\theta$	0	$\dfrac{1}{\sqrt{3}}$	1	$\sqrt{3}$	U.D.	0	U.D.	0
$Cot\,\theta$	U.D.	$\sqrt{3}$	1	$\dfrac{1}{\sqrt{3}}$	0	U.D.	0	U.D.
$Sec\,\theta$	1	$\dfrac{2}{\sqrt{3}}$	$\sqrt{2}$	2	U.D.	-1	U.D.	1
$Cosec\,\theta$	U.D.	2	$\sqrt{2}$	$\dfrac{2}{\sqrt{3}}$	1	U.D.	-1	U.D.

Trigonometric identities:

1. $Sin^2\theta + Cos^2\theta = 1$

2. $Sec^2\theta - Tan^2\theta = 1$

From this we can write,

$(Sec\theta + Tan\theta)(Sec\theta - Tan\theta) = 1$

$Sec\theta + Tan\theta = \dfrac{1}{Sec\theta - Tan\theta}$ or $Sec\theta - Tan\theta = \dfrac{1}{Sec\theta + Tan\theta}$

i.e. If $Sec\theta + Tan\theta = \dfrac{p}{q}$ then $Sec\theta - Tan\theta = \dfrac{q}{p}$

3. $Cosec^2\theta - Cot^2\theta = 1$

From this we can write,

$(Cosec\theta + Cot\theta)(Cosec\theta - Cot\theta) = 1$

$Cosec\theta + Cot\theta = \dfrac{1}{Cosec\theta - Cot\theta}$ or $Cosec\theta - Cot\theta = \dfrac{1}{Cosec\theta + Cot\theta}$

i.e. If $Cosec\theta + Cot\theta = \dfrac{p}{q}$ then $Cosec\theta - Cot\theta = \dfrac{q}{p}$

4. $Sin\ \theta . Cosec\ \theta = 1;\ Sec\ \theta . Cos\ \theta = 1;\ Tan\ \theta . Cot\ \theta = 1$

i.e. $Cosec\ \theta = \dfrac{1}{Sin\theta};\ Sec\ \theta = \dfrac{1}{Cos\ \theta};\ Cot\ \theta = \dfrac{1}{Tan\ \theta}$

Complimentary Angles:

The two angles A and B are said to be complementary angles when $A + B = 90^0$

Eg: $30^0\ and\ 60^0$

Supplementary Angles:

The two angles A and B are said to be supplementary angles when $A + B = 180^0$

Eg: $130^0\ and\ 50^0$

Note: If $A + B = 90^0$ then

(i) $Sin^2A + Sin^2B = 1$

(ii) $Cos^2A + Cos^2B = 1$

(iii) $Tan\ A . Tan\ B = 1$

(iv) $Cot\ A . Cot\ B = 1$

18. PERIODICITY AND EXTREME VALUES

Domain and Range of Trigonometric Functions:

Trigonometric Ratio	Domain	Range
Sin θ	$\mathbb{R}$	$[-1,1]$
Cos θ	$\mathbb{R}$	$[-1,1]$
Tan θ	$\mathbb{R} - \left\{(2n+1)\dfrac{\pi}{2}\right\} \forall n \in Z$	$\mathbb{R}$
Cot θ	$\mathbb{R} - \{n\pi\} \forall n \in Z$	$\mathbb{R}$
Sec θ	$\mathbb{R} - \left\{(2n+1)\dfrac{\pi}{2}\right\} \forall n \in Z$	$(-\infty, -1] \cup [1, \infty)$
Cosec θ	$\mathbb{R} - \{n\pi\} \forall n \in Z$	$(-\infty, -1] \cup [1, \infty)$

Period: Let $f(x)$ be a function. Then f is said to be a periodic function when $f(x + P) = f(x)$ where P is a least positive integer.

Here P is called the Period of the function.

The period of $f(ax \pm b)$ is $\dfrac{P}{|a|}$

The period of $f(x) = \dfrac{af_1(x) \pm bf_2(x)}{cf_3(x) \pm df_4(x)}$ is the L.C.M. of the periods of $f_1(x), f_2(x), f_3(x), f_4(x)$

The period of $|Sinx|, |Cosx|, |Tanx|, |Cotx|, |Secx|, |Cosecx|$ is π

The period of $|Sinx + Cosx|, |Sinx - Cosx|, |Cosx - Sinx|$ is π

The period of $|Tanx + Cotx|, |tanx - Cotx|, |Cotx - tanx|$ is $\dfrac{\pi}{2}$

The period of $Sin^{2n}x + Cos^{2n}x, Tan^{2n}x + Cot^{2n}x, Sec^{2n}x + Cosec^{2n}x$ is $\dfrac{\pi}{2}$ where $n \in Z^+$

The period of $aSin^{2n}x + bCos^{2m}x, aTan^{2n}x + bCot^{2m}x, aSec^{2n}x + bCosec^{2m}x$ is π where $m, n \in Z^+$

Trigonometric Function	Period
$\sin x$	2π
$\cos x$	2π
$\tan x$	π
$\cot x$	π
$\sec x$	2π
$\csc x$	2π
$\sin(ax \pm b)$	$\dfrac{2\pi}{|a|}$
$\cos(ax \pm b)$	$\dfrac{2\pi}{|a|}$
$\tan(ax \pm b)$	$\dfrac{\pi}{|a|}$
$\cot(ax \pm b)$	$\dfrac{\pi}{|a|}$
$\sec(ax \pm b)$	$\dfrac{2\pi}{|a|}$
$\csc(ax \pm b)$	$\dfrac{2\pi}{|a|}$

Graphs of Trigonometric Functions:

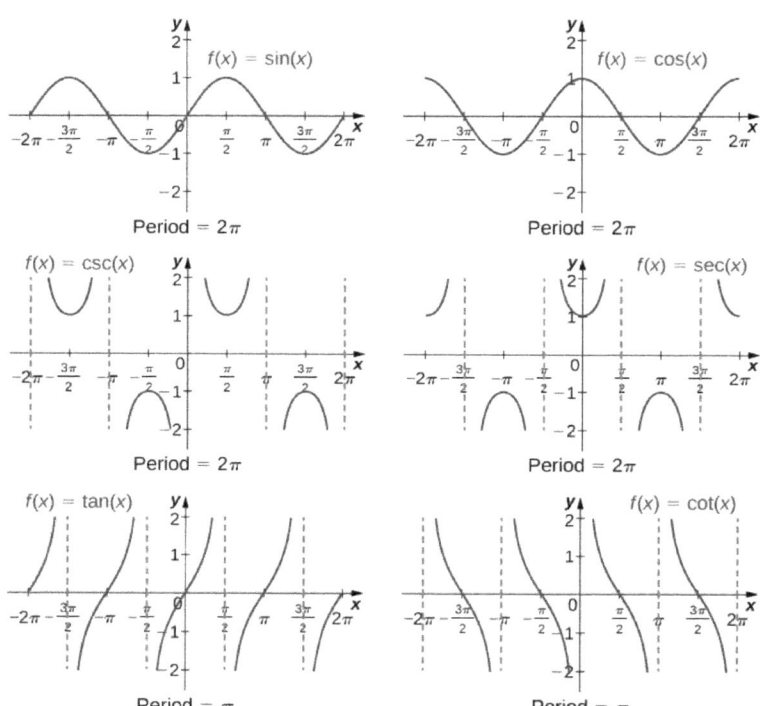

Extreme Values:

The extreme values of the function $f(x) = a\cos x + b\sin x + c$ are given by $c \pm \sqrt{a^2 + b^2}$

19. COMPOUND ANGLES

1. $Sin(A + B) = SinA\, CosB + CosA\, SinB$
2. $Sin(A - B) = SinA\, CosB - CosA\, SinB$
3. $Cos(A + B) = CosA\, CosB - SinA\, SinB$
4. $Cos(A - B) = CosA\, CosB + SinA\, SinB$
5. $Tan(A + B) = \dfrac{TanA + TanB}{1 - TanA\, TanB}$
6. $Tan(A - B) = \dfrac{TanA - TanB}{1 + TanA\, TanB}$
7. $Cot(A + B) = \dfrac{CotA\, CotB - 1}{CotA + CotB}$
8. $Cot(A - B) = \dfrac{CotA\, CotB + 1}{CotB - CotA}$
9. $Sin(A + B) + Sin(A - B) = 2\, SinA\, CosB$
10. $Sin(A + B) - Sin(A - B) = 2\, CosA\, SinB$
11. $Cos(A + B) + Cos(A - B) = 2\, CosA\, CosB$
12. $Cos(A + B) - Cos(A - B) = -2\, SinA\, SinB$ or
    $Cos(A - B) - Cos(A + B) = 2\, SinA\, SinB$
13. $Sin(A + B).Sin(A - B) = Sin^2A - Sin^2B = Cos^2B - Cos^2A$
14. $Cos(A + B).Cos(A - B) = Cos^2A - Sin^2B = Cos^2B - Sin^2A$
15. $Tan(A + B).Tan(A - B) = \dfrac{Tan^2A - Tan^2B}{1 - Tan^2A\, Tan^2B}$
16. $Cot(A + B).Cot(A - B) = \dfrac{Cot^2A - Cot^2B}{1 - Cot^2A\, Cot^2B}$
17. $Tan(45^0 + A) = \dfrac{1 + TanA}{1 - TanA} = Cot(45^0 - A)$
18. $Tan(45^0 - A) = \dfrac{1 - TanA}{1 + TanA} = Cot(45^0 + A)$
19. If $A + B = 45^0\ or\ 225^0$ then
    (i) $(1 + TanA)(1 + TanB) = 2$
    (ii) $(1 - CotA)(1 - CotB) = 2$
    (iii) $(1 + CotA)(1 + CotB) = 2\, CotA\, CotB$
20. If $A + B = 135^0\ or\ 315^0$ then
    (i) $(1 - TanA)(1 - TanB) = 2$
    (ii) $(1 + CotA)(1 + CotB) = 2$
    (iii) $(1 + TanA)(1 + TanB) = 2\, TanA\, TanB$

21. $Sin(A + B + C) = SinA\, CosB\, CosC + CosA\, SinB\, CosC + CosA\, CosB\, SinC - SinA\, SinB\, SinC$

22. $Cos(A + B + C) = CosA\, CosB\, CosC - CosA\, SinB\, SinC - SinA\, CosB\, SinC - SinA\, SinB\, CosC$

23. $Tan(A + B + C) = \dfrac{TanA + TanB + TanC - TanA\, TanB\, TanC}{1 - TanA\, TanB - TanB\, TanC - TanC\, TanA} = \dfrac{\sum TanA - \prod TanA}{1 - \sum TanA\, TanB}$

24. $Tan(A + B + C) = \dfrac{CotA + CotB + CotC - CotA\, CotB\, CotC}{1 - CotA\, CotB - CotB\, CotC - CotC\, CotA} = \dfrac{\sum CotA - \prod CotA}{1 - \sum CotA\, CotB}$

25. If $A + B + C = n\pi\ \forall n \in Z$ then

    (i) $\sum TanA = \prod TanA$ or $TanA + TanB + TanC = TanA\, TanB\, TanC$

    (ii) $\sum CotA\, CotB = 1$ or $CotA\, CotB + CotB\, CotC + CotC\, CotA = 1$

26. If $A + B + C = (2n + 1)\dfrac{\pi}{2}\ \forall n \in Z$ then

    (i) $\sum CotA = \prod CotA$ (or) $CotA + CotB + CotC = CotA\, CotB\, CotC$

    (ii) $\sum TanA\, TanB = 1$ (or) $TanA\, TanB + TanB\, TanC + TanC\, TanA = 1$

27.

	15^0	75^0
Sin	$\dfrac{\sqrt{3} - 1}{2\sqrt{2}}$	$\dfrac{\sqrt{3} + 1}{2\sqrt{2}}$
Cos	$\dfrac{\sqrt{3} + 1}{2\sqrt{2}}$	$\dfrac{\sqrt{3} - 1}{2\sqrt{2}}$
Tan	$2 - \sqrt{3}$	$2 + \sqrt{3}$
Cot	$2 + \sqrt{3}$	$2 - \sqrt{3}$
Sec	$\sqrt{6} - \sqrt{2}$	$\sqrt{6} + \sqrt{2}$
Cosec	$\sqrt{6} + \sqrt{2}$	$\sqrt{6} - \sqrt{2}$

20. MULTIPLE AND SUB-MULTIPLE ANGLES

1. $Sin2A = 2SinA\, CosA = \dfrac{2\,TanA}{1+Tan^2 A}$

2. $Cos2A = Cos^2 A - Sin^2 A = 2Cos^2 A - 1 = 1 - 2Sin^2 A = \dfrac{1-Tan^2 A}{1+Tan^2 A}$

3. $Tan2A = \dfrac{2\,TanA}{1-Tan^2 A}$

4. $Cot2A = \dfrac{Cot^2 A - 1}{2\,CotA}$

5. $SinA = 2Sin\dfrac{A}{2} Cos\dfrac{A}{2} = \dfrac{2\,Tan\frac{A}{2}}{1+Tan^2\frac{A}{2}}$

6. $CosA = Cos^2\dfrac{A}{2} - Sin^2\dfrac{A}{2} = 2Cos^2\dfrac{A}{2} - 1 = 1 - 2Sin^2\dfrac{A}{2} = \dfrac{1-Tan^2\frac{A}{2}}{1+Tan^2\frac{A}{2}}$

7. $TanA = \dfrac{2\,Tan\frac{A}{2}}{1-Tan^2\frac{A}{2}}$

8. $Cot2A = \dfrac{Cot^2\frac{A}{2} - 1}{2\,Cot\frac{A}{2}}$

9. $SinA = \pm\sqrt{\dfrac{1-cos2A}{2}}$

10. $CosA = \pm\sqrt{\dfrac{1+cos2A}{2}}$

11. $TanA = \pm\sqrt{\dfrac{1-cos2A}{1+Cos2A}}$

12. $CotA = \pm\sqrt{\dfrac{1+cos2A}{1-Cos2A}}$

13. $Sin\dfrac{A}{2} = \pm\sqrt{\dfrac{1-cosA}{2}}$

14. $Cos\dfrac{A}{2} = \pm\sqrt{\dfrac{1+cosA}{2}}$

15. $Tan\dfrac{A}{2} = \pm\sqrt{\dfrac{1-cosA}{1+CosA}}$

16. $Cot\dfrac{A}{2} = \pm\sqrt{\dfrac{1+cosA}{1-CosA}}$

17. $Sin3A = 3SinA - 4Sin^3 A$

18. $Cos3A = 4Cos^3 A - 3CosA$

19. $Tan3A = \dfrac{3TanA - Tan^3 A}{1 - 3Tan^2 A}$

20. $Cot3A = \dfrac{3CotA - Cot^3A}{1 - 3Cot^2A}$

21. $CotA + TanA = 2Cosec\,2A$

22. $CotA - TanA = 2Cot\,2A$

23. $TanA + 2Tan2A + 4Tan4A + \cdots\ldots + 2^{n-1}Tan2^{n-1}A + 2^n Cot2^nA = CotA$

24. $CosA.Cos2A.Cos2^2A.Cos2^3A \ldots\ldots\ldots\ldots Cos2^{n-1}A = \dfrac{Sin2^nA}{2^n SinA}$

25. If $A + B = 60^0$ then

   (i) $Sin^2A + Sin^2B + SinA.SinB = \dfrac{3}{4}$

   (ii) $Cos^2A + Cos^2B + CosA.CosB = \dfrac{3}{4}$

26. If $A - B = 60^0$ then

   (i) $Sin^2A + Sin^2B - SinA.SinB = \dfrac{3}{4}$

   (ii) $Cos^2A + Cos^2B - CosA.CosB = \dfrac{3}{4}$

27. If $\alpha = 60^0\ or\ 120^0\ or\ 240^0\ or\ 300^0$ then

   (i) $Sin\theta.Sin(\theta - \alpha).Sin(\theta + \alpha) = \dfrac{1}{4}Sin3\theta$

   (ii) $Cos\theta.Cos(\theta - \alpha).Cos(\theta + \alpha) = \dfrac{1}{4}Cos3\theta$

   (iii) $Tan\theta.Tan(\theta - \alpha).Tan(\theta + \alpha) = Tan3\theta$

28. Let $C = Cos\dfrac{A}{2}$ and $S = Sin\dfrac{A}{2}$ then $C + S = \pm\sqrt{1 + SinA}$ and $C - S = \pm\sqrt{1 - SinA}$

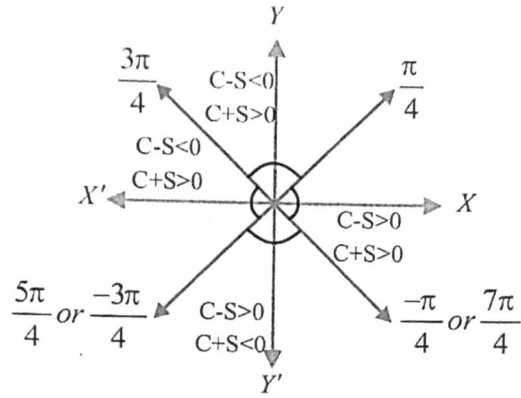

   (i) $C + S > 0$ in $\left(-\dfrac{\pi}{4}, \dfrac{3\pi}{4}\right)$

   (ii) $C + S < 0$ in $\left(\dfrac{3\pi}{4}, \dfrac{7\pi}{4}\right)$

   (iii) $C - S > 0$ in $\left(-\dfrac{3\pi}{4}, \dfrac{\pi}{4}\right)$

   (iv) $C - S < 0$ in $\left(\dfrac{\pi}{4}, \dfrac{5\pi}{4}\right)$

29. (i) $\sqrt{2+\sqrt{2+\sqrt{2+\cdots\ldots\sqrt{2+2Cos\theta}}}} = 2Cos\left(\dfrac{\theta}{2^n}\right)$ where 'n' is the number of square roots and $0 < \theta < \pi$

(ii) $\sqrt{2+\sqrt{2+\sqrt{2+\cdots\ldots}}} = 2Cos\left(\dfrac{\pi}{2^{n+1}}\right)$ where 'n' is the number of square roots

30.

	$18°$	$36°$	$54°$	$72°$
Sin	$\dfrac{\sqrt{5}-1}{4}$	$\dfrac{\sqrt{10-2\sqrt{5}}}{4}$	$\dfrac{\sqrt{5}+1}{4}$	$\dfrac{\sqrt{10+2\sqrt{5}}}{4}$
Cos	$\dfrac{\sqrt{10+2\sqrt{5}}}{4}$	$\dfrac{\sqrt{5}+1}{4}$	$\dfrac{\sqrt{10-2\sqrt{5}}}{4}$	$\dfrac{\sqrt{5}-1}{4}$

	$22\dfrac{1}{2}°$	$67\dfrac{1}{2}°$
Sin	$\dfrac{\sqrt{2-\sqrt{2}}}{2}$	$\dfrac{\sqrt{2+\sqrt{2}}}{2}$
Cos	$\dfrac{\sqrt{2+\sqrt{2}}}{2}$	$\dfrac{\sqrt{2-\sqrt{2}}}{2}$
Tan	$\sqrt{2}-1$	$\sqrt{2}+1$
Cot	$\sqrt{2}+1$	$\sqrt{2}-1$

21. TRANSFORMATIONS

1. $SinC + SinD = 2Sin\left(\frac{C+D}{2}\right)Cos\left(\frac{C-D}{2}\right)$

2. $SinC - SinD = 2Cos\left(\frac{C+D}{2}\right)Sin\left(\frac{C-D}{2}\right)$

3. $CosC + CosD = 2Cos\left(\frac{C+D}{2}\right)Cos\left(\frac{C-D}{2}\right)$

4. $CosC - CosD = -2Sin\left(\frac{C+D}{2}\right)Sin\left(\frac{C-D}{2}\right)$

(or) $CosD - CosC = 2Sin\left(\frac{C+D}{2}\right)Sin\left(\frac{C-D}{2}\right)$

22. TRIGONOMETRIC EQUATIONS

S.No.	Trigonometric Equation	Principal Value	General Solution
1	$Sin\theta = 0$	0^0	$\theta = n\pi \quad \forall n \in Z$
2	$Cos\theta = 0$	$\dfrac{\pi}{2}$	$\theta = (2n+1)\dfrac{\pi}{2} \quad \forall n \in Z$
3	$Tan\theta = 0$	0^0	$\theta = n\pi \quad \forall n \in Z$
4	$Sin\theta = Sin\alpha$	$\alpha \in \left[-\dfrac{\pi}{2}, \dfrac{\pi}{2}\right]$	$\theta = n\pi + (-1)^n \alpha \quad \forall n \in Z$
5	$Cos\theta = Cos\alpha$	$\alpha \in [0, \pi]$	$\theta = 2n\pi \pm \alpha \quad \forall n \in Z$
6	$Tan\theta = Tan\alpha$	$\alpha \in \left(-\dfrac{\pi}{2}, \dfrac{\pi}{2}\right)$	$\theta = n\pi + \alpha \quad \forall n \in Z$
7	$Sin^2\theta = Sin^2\alpha$ $Cos^2\theta = Cos^2\alpha$ $Tan^2\theta = Tan^2\alpha$	$\alpha \in \left(0, \dfrac{\pi}{2}\right)$	$\theta = n\pi \pm \alpha \quad \forall n \in Z$

The trigonometric equation $aCos\theta + bSin\theta = c$ possesses

(i) A solution when $|c| \leq \sqrt{a^2 + b^2}$

(ii) No solution when $|c| > \sqrt{a^2 + b^2}$

23. INVERSE TRIGONOMETRIC FUNCTIONS

Inverse trigonometric Function	Domain	Range
$Sin^{-1}x$	$[-1,1]$	$\left[-\dfrac{\pi}{2},\dfrac{\pi}{2}\right]$
$Cos^{-1}x$	$[-1,1]$	$[0,\pi]$
$Tan^{-1}x$	$\mathbb{R}$	$\left(-\dfrac{\pi}{2},\dfrac{\pi}{2}\right)$
$Cot^{-1}x$	$\mathbb{R}$	$(0,\pi)$
$Sec^{-1}x$	$(-\infty,-1]\cup[1,\infty)$	$[0,\dfrac{\pi}{2})\cup(\dfrac{\pi}{2},\pi]$
$Cosec^{-1}x$	$(-\infty,-1]\cup[1,\infty)$	$[-\dfrac{\pi}{2},0)\cup(0,\dfrac{\pi}{2}]$

Graphs of Inverse Trigonometric Functions:

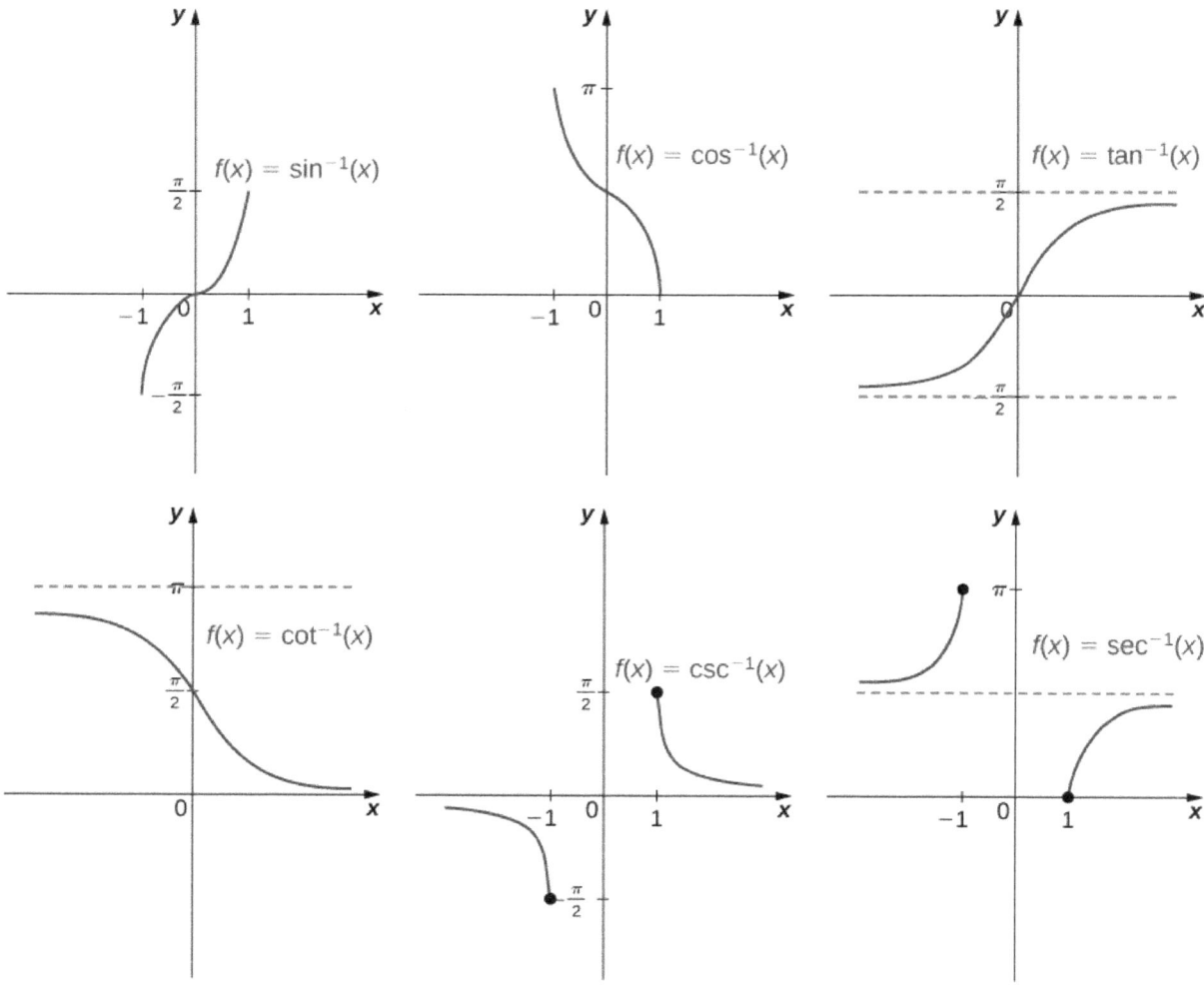

MATHEMATICS SUCCESS MANTRA

1. (i) $Sin^{-1}(x) = Cosec^{-1}\left(\frac{1}{x}\right) \ \forall x \in [-1,1]$ and $x \neq 0$

(ii) $Cos^{-1}(x) = Sec^{-1}\left(\frac{1}{x}\right) \ \forall x \in [-1,1]$ and $x \neq 0$

(iii) $Tan^{-1}(x) = \begin{cases} Cot^{-1}\left(\frac{1}{x}\right) & \forall x > 0 \\ -\pi + Cot^{-1}\left(\frac{1}{x}\right) & \forall x < 0 \end{cases}$

2. (i) $Sin^{-1}(Sinx) = x \ \ \forall x \in \left[-\frac{\pi}{2},\frac{\pi}{2}\right]$

(ii) $Cos^{-1}(Cosx) = x \ \ \forall x \in [0,\pi]$

(iii) $Tan^{-1}(Tanx) = x \ \ \forall x \in \left(-\frac{\pi}{2},\frac{\pi}{2}\right)$

(iv) $Cot^{-1}(Cotx) = x \ \ \forall x \in (0,\pi)$

(v) $Sec^{-1}(Secx) = x \ \ \forall x \in [0,\frac{\pi}{2}) \cup (\frac{\pi}{2},\pi]$

(vi) $Cosec^{-1}(Cosecx) = x \ \ \forall x \in [-\frac{\pi}{2},0) \cup (0,\frac{\pi}{2}]$

3. (i) $Sin(Sin^{-1}x) = x \ \ \forall x \in [-1,1]$

(ii) $Cos(Cos^{-1}x) = x \ \ \forall x \in [-1,1]$

(iii) $Tan(Tan^{-1}x) = x \ \ \forall x \in \mathbb{R}$

(iv) $Cot(Cot^{-1}x) = x \ \ \forall x \in \mathbb{R}$

(v) $Sec(Sec^{-1}x) = x \ \ \forall x \in (-\infty,-1] \cup [1,\infty)$

(vi) $Cosec(Cosec^{-1}x) = x \ \ \forall x \in (-\infty,-1] \cup [1,\infty)$

4. (i) $Sin^{-1}(-x) = -Sin^{-1}(x) \ \forall x \in [-1,1]$

(ii) $Cos^{-1}(-x) = \pi - Cos^{-1}(x) \ \forall x \in [-1,1]$

(iii) $Tan^{-1}(-x) = -Tan^{-1}(x) \ \forall x \in \mathbb{R}$

(iv) $Cot^{-1}(-x) = \pi - Cot^{-1}(x) \ \forall x \in \mathbb{R}$

(v) $Sec^{-1}(-x) = \pi - Sec^{-1}(x) \ \forall x \in (-\infty,-1] \cup [1,\infty)$

(vi) $Cosec^{-1}(-x) = -Cosec^{-1}(x) \ \forall x \in (-\infty,-1] \cup [1,\infty)$

5. (i) $Sin^{-1}x + Cos^{-1}x = \frac{\pi}{2} \ \ \forall x \in [-1,1]$

(ii) $Tan^{-1}x + Cot^{-1}x = \frac{\pi}{2} \ \ \forall x \in \mathbb{R}$

(iii) $Sec^{-1}x + Cosec^{-1}x = \frac{\pi}{2} \ \ \forall x \in (-\infty,-1] \cup [1,\infty)$

6. (i) $Sin^{-1}x \pm Sin^{-1}y = Sin^{-1}(x\sqrt{1-y^2} \pm y\sqrt{1-x^2})$

$\forall x, y \in [-1,1], xy < 0 \text{ and } x^2 + y^2 > 1$

(ii) $Cos^{-1}x \pm Cos^{-1}y = Cos^{-1}(xy \mp \sqrt{1-x^2}\sqrt{1-y^2})$ $\forall x, y \in [-1,1], x + y \geq 0$

(iii) $Tan^{-1}(x) + Tan^{-1}(y) = \begin{cases} Tan^{-1}\left(\frac{x+y}{1-xy}\right) & \text{if } x > 0, y > 0, xy < 1 \\ \pi + Tan^{-1}\left(\frac{x+y}{1-xy}\right) & \text{if } x > 0, y > 0, xy > 1 \\ -\pi + Tan^{-1}\left(\frac{x+y}{1-xy}\right) & \text{if } x < 0, y < 0, xy > 1 \end{cases}$

(iv) $Tan^{-1}(x) - Tan^{-1}(y) = \begin{cases} Tan^{-1}\left(\frac{x-y}{1+xy}\right) & \text{if } xy > -1 \\ \pi + Tan^{-1}\left(\frac{x+y}{1-xy}\right) & \text{if } x > 0, y < 0, xy < -1 \\ -\pi + Tan^{-1}\left(\frac{x+y}{1-xy}\right) & \text{if } x < 0, y > 0, xy < -1 \end{cases}$

7. (i) $2Sin^{-1}x = Sin^{-1}(2x\sqrt{1-x^2})$ if $x \in \left[\frac{-1}{\sqrt{2}}, \frac{1}{\sqrt{2}}\right]$

(ii) $2Cos^{-1}x = \begin{cases} Cos^{-1}(2x^2 - 1) & \text{if } x \in [0,1] \\ 2\pi - Cos^{-1}(2x^2 - 1) & \text{if } x \in [-1,0] \end{cases}$

(iii) $2Tan^{-1}x = \begin{cases} Sin^{-1}\left(\frac{2x}{1+x^2}\right) & \text{if } x \in [-1,1] \\ Cos^{-1}\left(\frac{1-x^2}{1+x^2}\right) & \text{if } x \geq 0 \\ Tan^{-1}\left(\frac{2x}{1-x^2}\right) & \text{if } x \in (-1,1) \end{cases}$

8. (i) $3Sin^{-1}x = Sin^{-1}(3x - 4x^3)$ if $x \in \left[-\frac{1}{2}, \frac{1}{2}\right]$

(ii) $3Cos^{-1}x = Cos^{-1}(4x^3 - 3x)$ if $x \in \left[\frac{1}{2}, 1\right]$

(iii) $3Tan^{-1}x = Tan^{-1}\left(\frac{3x-x^3}{1-3x^2}\right)$ if $x \in \left(-\frac{1}{\sqrt{3}}, \frac{1}{\sqrt{3}}\right)$

9. (i) $Tan^{-1}x + Tan^{-1}y + Tan^{-1}z = Tan^{-1}\left(\frac{x+y+z-xyz}{1-xy-yz-zx}\right)$

(ii) If $Tan^{-1}x + Tan^{-1}y + Tan^{-1}z = \frac{\pi}{2}$ then $xy + yz + zx = 1$

(iii) If $Tan^{-1}x + Tan^{-1}y + Tan^{-1}z = \pi$ then $x + y + z = xyz$

24. PROPERTIES OF TRIANGLES

Representation of sides of a triangle:

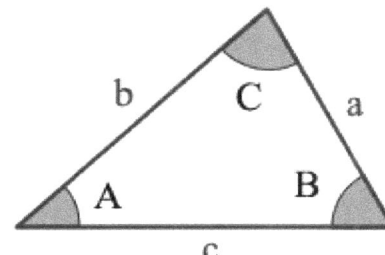

Let $\triangle ABC$ be a triangle. Then the sides opposite to the angles A, B, C are BC, CA, AB are represented by the letters a, b, c.

Semi perimeter: The perimeter of the triangle is $a + b + c$.

The semi perimeter is represented by s and is given by $s = \dfrac{a+b+c}{2}$

Radii of circles:

R is the radius of circumcircle.

r is the radius of incircle.

r_1, r_2, r_3 are the radii of excircles.

Sine Rule: $\dfrac{a}{SinA} = \dfrac{b}{SinB} = \dfrac{c}{SinC} = 2R$ where R is the circumradius of $\triangle ABC$

i.e. $a = 2R SinA; b = 2R SinB; c = 2R SinC$

Cosine Rule:

(i) $a^2 = b^2 + c^2 - 2bc\, CosA$

(ii) $b^2 = c^2 + a^2 - 2ca\, CosB$

(iii) $c^2 = a^2 + b^2 - 2ab\, CosC$

i.e. $CosA = \dfrac{b^2+c^2-a^2}{2bc}$; $CosB = \dfrac{c^2+a^2-b^2}{2ca}$; $CosC = \dfrac{a^2+b^2-c^2}{2ab}$

Tangent Rule (Napier's Rule):

(i) $Tan\left(\dfrac{A-B}{2}\right) = \dfrac{a-b}{a+b} Cot\dfrac{C}{2} = \dfrac{a-b}{a+b} Tan\left(\dfrac{A+B}{2}\right)$

(ii) $Tan\left(\dfrac{B-C}{2}\right) = \dfrac{b-c}{b+c} Cot\dfrac{A}{2} = \dfrac{b-c}{b+c} Tan\left(\dfrac{B+C}{2}\right)$

(iii) $Tan\left(\dfrac{C-A}{2}\right) = \dfrac{c-a}{c+a} Cot\dfrac{B}{2} = \dfrac{c-a}{c+a} Tan\left(\dfrac{C+A}{2}\right)$

Mollweide Rule:

(i) $Sin\left(\frac{A-B}{2}\right) = \frac{a-b}{c} Cos\frac{C}{2}$; $Sin\left(\frac{B-C}{2}\right) = \frac{b-c}{a} Cos\frac{A}{2}$; $Sin\left(\frac{C-A}{2}\right) = \frac{c-a}{b} Cos\frac{B}{2}$

(ii) $Cos\left(\frac{A-B}{2}\right) = \frac{a+b}{c} Sin\frac{C}{2}$; $Cos\left(\frac{B-C}{2}\right) = \frac{b+c}{a} Sin\frac{A}{2}$; $Cos\left(\frac{C-A}{2}\right) = \frac{c+a}{b} Sin\frac{B}{2}$

Projection Rule:

(i) $a = b\, CosC + c\, CosB$

(ii) $b = a\, CosC + c\, CosA$

(iii) $c = a\, CosB + b\, CosA$

Area of Triangle: The area of triangle is represented by $\triangle$

(i) $\triangle = \frac{1}{2} bc\, SinA = \frac{1}{2} ca\, SinB = \frac{1}{2} ab\, SinC$

(ii) $\triangle = \sqrt{s(s-a)(s-b)(s-c)}$

(iii) $\triangle = \frac{abc}{4R}$

(iv) $\triangle = 2R^2 SinA\, SinB\, SinC$

(v) $\triangle = rs$

(vi) $\triangle = \sqrt{r \cdot r_1 \cdot r_2 \cdot r_3}$

Half-Angle Formulae:

(i) $Sin\frac{A}{2} = \sqrt{\frac{(s-b)(s-c)}{bc}}$; $Sin\frac{B}{2} = \sqrt{\frac{(s-c)(s-a)}{ca}}$; $Sin\frac{C}{2} = \sqrt{\frac{(s-a)(s-b)}{ab}}$

(ii) $Cos\frac{A}{2} = \sqrt{\frac{s(s-a)}{bc}}$; $Cos\frac{B}{2} = \sqrt{\frac{s(s-b)}{ac}}$; $Cos\frac{C}{2} = \sqrt{\frac{s(s-c)}{ab}}$

(iii) $Tan\frac{A}{2} = \sqrt{\frac{(s-b)(s-c)}{s(s-a)}} = \frac{(s-b)(s-c)}{\triangle} = \frac{\triangle}{s(s-a)}$

$Tan\frac{B}{2} = \sqrt{\frac{(s-c)(s-a)}{s(s-b)}} = \frac{(s-c)(s-a)}{\triangle} = \frac{\triangle}{s(s-b)}$

$Tan\frac{C}{2} = \sqrt{\frac{(s-a)(s-b)}{s(s-c)}} = \frac{(s-a)(s-b)}{\triangle} = \frac{\triangle}{s(s-c)}$

(iv) $Cot\frac{A}{2} = \sqrt{\frac{s(s-a)}{(s-b)(s-c)}} = \frac{s(s-a)}{\triangle} = \frac{\triangle}{(s-b)(s-c)}$

$Cot\frac{B}{2} = \sqrt{\frac{s(s-b)}{(s-c)(s-a)}} = \frac{s(s-b)}{\triangle} = \frac{\triangle}{(s-c)(s-a)}$

$Cot\frac{C}{2} = \sqrt{\frac{s(s-c)}{(s-a)(s-b)}} = \frac{s(s-c)}{\triangle} = \frac{\triangle}{(s-a)(s-b)}$

Inradius:

(i) $r = \dfrac{\Delta}{s}$

(ii) $r = 4R \sin\dfrac{A}{2} \sin\dfrac{B}{2} \sin\dfrac{C}{2}$

(iii) $r = (s-a)\tan\dfrac{A}{2} = (s-b)\tan\dfrac{B}{2} = (s-c)\tan\dfrac{C}{2}$

(iv) $r = \dfrac{a}{\cot\dfrac{B}{2}+\cot\dfrac{C}{2}} = \dfrac{b}{\cot\dfrac{C}{2}+\cot\dfrac{A}{2}} = \dfrac{c}{\cot\dfrac{A}{2}+\cot\dfrac{B}{2}}$

Ex-Radii:

(i) $r_1 = \dfrac{\Delta}{s-a}$; $r_2 = \dfrac{\Delta}{s-b}$; $r_3 = \dfrac{\Delta}{s-c}$

(ii) $r_1 = s\tan\dfrac{A}{2} = (s-b)\cot\dfrac{C}{2} = (s-c)\cot\dfrac{B}{2}$

$r_2 = s\tan\dfrac{B}{2} = (s-a)\cot\dfrac{C}{2} = (s-c)\cot\dfrac{A}{2}$

$r_3 = s\tan\dfrac{C}{2} = (s-a)\cot\dfrac{B}{2} = (s-b)\cot\dfrac{A}{2}$

(iii) $r_1 = 4R \sin\dfrac{A}{2} \cos\dfrac{B}{2} \cos\dfrac{C}{2}$

$r_2 = 4R \cos\dfrac{A}{2} \sin\dfrac{B}{2} \cos\dfrac{C}{2}$

$r_3 = 4R \cos\dfrac{A}{2} \cos\dfrac{B}{2} \sin\dfrac{C}{2}$

(iv) $r_1 = \dfrac{a}{\tan\dfrac{B}{2}+\tan\dfrac{C}{2}}$; $r_2 = \dfrac{b}{\tan\dfrac{C}{2}+\tan\dfrac{A}{2}}$; $r_3 = \dfrac{c}{\tan\dfrac{A}{2}+\tan\dfrac{B}{2}}$

Relation between inradius, circumradius and ex-radii:

(i) $\dfrac{1}{r_1} + \dfrac{1}{r_2} + \dfrac{1}{r_3} = \dfrac{1}{r}$

(ii) $r_1 r_2 + r_2 r_3 + r_3 r_1 = s^2$

(iii) $r(r_1 + r_2 + r_3) = ab + bc + ca - s^2$

(iv) $r_1 + r_2 + r_3 - r = 4R$

$r + r_2 + r_3 - r_1 = 4R\cos A$

$r + r_1 + r_3 - r_2 = 4R\cos B$

$r + r_1 + r_2 - r_3 = 4R\cos C$

(v) $r^2 + r_1^2 + r_2^2 + r_3^2 = 16R^2 - (a^2 + b^2 + c^2)$

(vi) $\dfrac{1}{r^2} + \dfrac{1}{r_1^2} + \dfrac{1}{r_2^2} + \dfrac{1}{r_3^2} = \dfrac{a^2+b^2+c^2}{\Delta^2}$

(vii) $\cos A + \cos B + \cos C = 1 + \dfrac{r}{R}$

(viii) $\sin A + \sin B + \sin C = \dfrac{s}{R}$

Perpendiculars:

Let P_1, P_2, P_3 be the perpendiculars drawn from the vertices A, B, C of $\triangle ABC$ to the opposite sides. Then

(i) $\dfrac{1}{P_1} + \dfrac{1}{P_2} + \dfrac{1}{P_3} = \dfrac{1}{r}$

(ii) $\dfrac{1}{P_2} + \dfrac{1}{P_3} - \dfrac{1}{P_3} = \dfrac{1}{r_1}$; $\dfrac{1}{P_1} + \dfrac{1}{P_3} - \dfrac{1}{P_2} = \dfrac{1}{r_2}$; $\dfrac{1}{P_1} + \dfrac{1}{P_2} - \dfrac{1}{P_3} = \dfrac{1}{r_3}$

(iii) $\dfrac{CosA}{P_1} + \dfrac{CosB}{P_2} + \dfrac{CosC}{P_3} = \dfrac{1}{R}$

(iv) $P_1 P_2 P_3 = \dfrac{a^2 b^2 c^2}{8R^3} = \dfrac{8\Delta^3}{abc}$

Length of Medians:

Let AD, BE, CF be the medians drawn from the vertices A, B, C of $\triangle ABC$ to the opposite sides. Then

(i) $AD = \dfrac{1}{2}\sqrt{2b^2 + 2c^2 - a^2}$

(ii) $BE = \dfrac{1}{2}\sqrt{2c^2 + 2a^2 - b^2}$

(iii) $CF = \dfrac{1}{2}\sqrt{2a^2 + 2b^2 - c^2}$

25. HEIGHTS AND DISTANCES

Angle of Elevation: If the position of the object is above the position of the observation, then the angle made by the line joining object and observation point with the horizontal line drawn at the observation point is called as an angle of elevation.

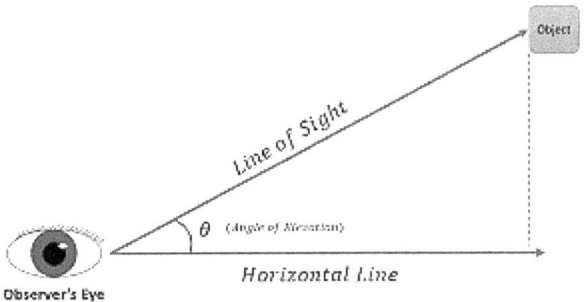

Angle of Depression: If the position of the object is below the position of the observation then the angle made by the line joining object and observation point with the horizontal line drawn at the observation point is called as an angle of depression.

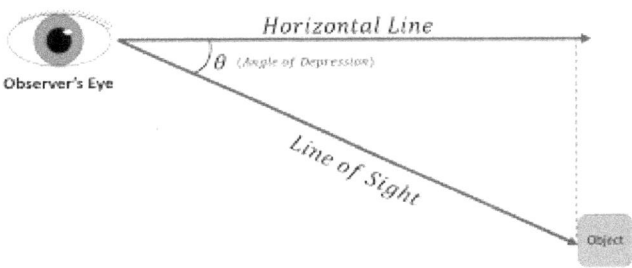

Directions:

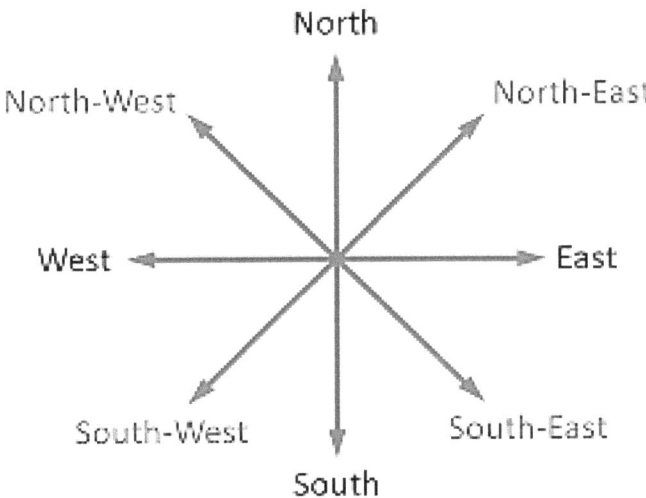

Standard diagrams of important models:

(i) The angle of elevation of the top of a tower of height h meters, standing on a horizontal plane, from a point is α. After walking a distance 'd' meters towards the foot of the tower, the angle of elevation is found to be β.

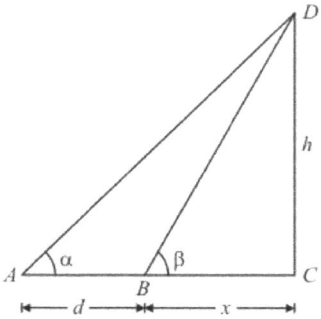

(ii) The observation points lie on either side of the tower of height h meters makes angles with top of the tower are α and β (or)

The angles of depression of two points on the level ground on either side of the observation point which is in the height of h from level ground are α and β

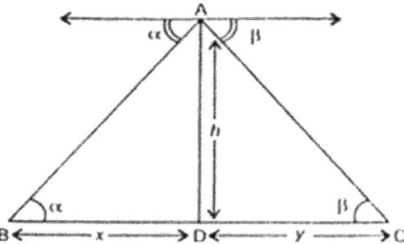

(iii) The angles of elevation of the top a tower of h meters from the top and bottom of a building of height H meters are α and β respectively

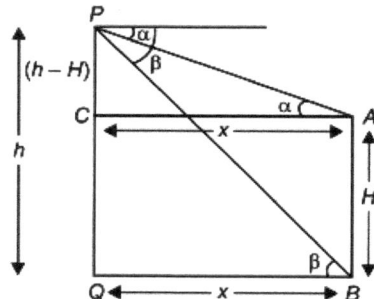

(iv) The angles of elevation of the tops of two towers observed from the midpoint on the line joining the bottoms of two towers of height H and h meters are α and β respectively

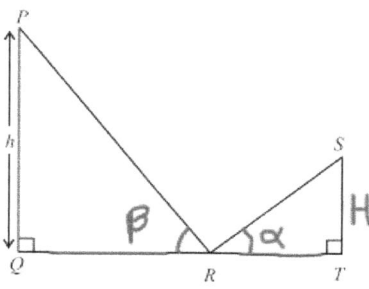

(v) The angles of elevation of the tops of two towers of same height observed from the any point on the midway of line joining the bottoms of two towers are α and β.

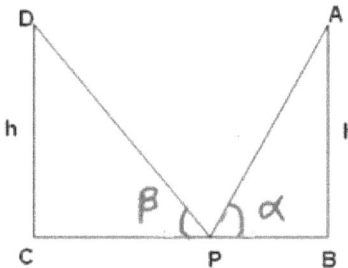

(v) The angles of elevation of top of two towers of height H and h meters observed from the bottoms of the other towers are α and β respectively

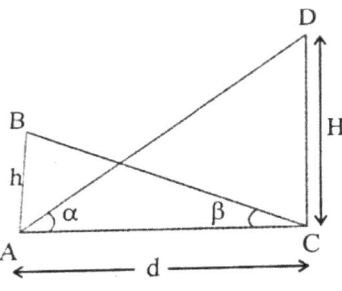

(vi) The angle of elevation and the angle of depression of a cloud is at a height of x meters from level ground and its image in the river observed from a point is at height of h meters from the level ground opposite to the river are α and β respectively

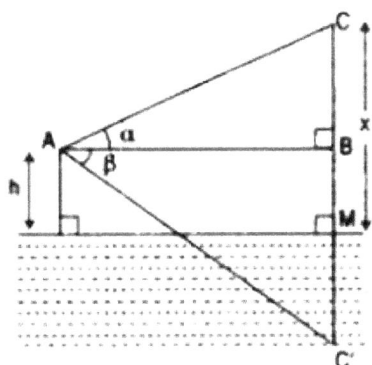

(vii) The angle of elevation and the angle of depression of the top and bottom of a tower of H meters observed from the the top of a tower of h meters are α and β respectively

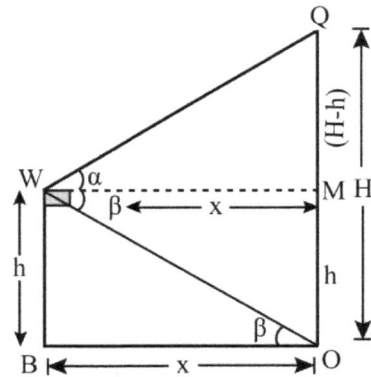

(viii) A ladder rests against a wall of height h meters at an angle α to the horizontal. Its foot is pulled away from the wall through a distance a meters so that it slides s distance b meters down the wall making an angle β with the horizontal

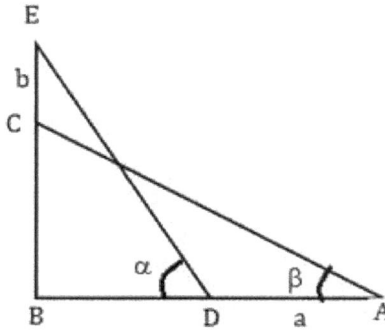

(ix) The angle of elevation of a hill of height h meters from a point is α. After walking to some point at a distance a meters from first point on a slope inclined at γ to the horizon, the angle of elevation was found to be β.

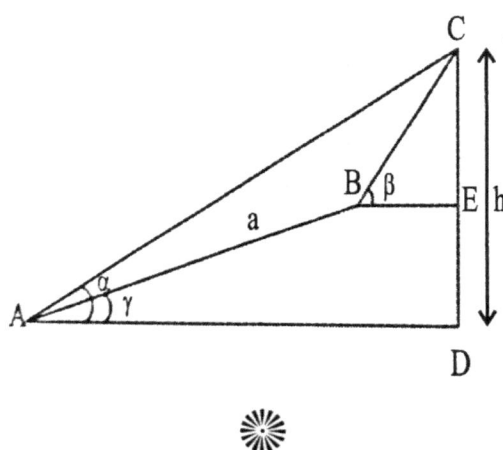

26. HYPERBOLIC FUNCTIONS

Expansions:

(i) $e^x = 1 + \frac{x}{1!} + \frac{x^2}{2!} + \frac{x^3}{3!} + \cdots \ldots \ldots \infty$

(ii) $e^{-x} = 1 - \frac{x}{1!} + \frac{x^2}{2!} - \frac{x^3}{3!} + \cdots \ldots \ldots \infty$

Hyperbolic Functions:

(i) $Sinhx = \frac{e^x - e^{-x}}{2}$ (ii) $Coshx = \frac{e^x + e^{-x}}{2}$ (iii) $Tanhx = \frac{e^x - e^{-x}}{e^x + e^{-x}}$

(iv) $Cothx = \frac{e^x + e^{-x}}{e^x - e^{-x}}$ (v) $Sechx = \frac{2}{e^x + e^{-x}}$ (iv) $Cosechx = \frac{2}{e^x - e^{-x}}$

Hyperbolic Function	Domain	Range
$Sinhx$	$\mathbb{R}$	$\mathbb{R}$
$Coshx$	$\mathbb{R}$	$[1, \infty)$
$Tanhx$	$\mathbb{R}$	$(-1, 1)$
$Cothx$	$\mathbb{R} - \{0\}$	$\mathbb{R} - [-1, 1]$
$Sechx$	$\mathbb{R}$	$(0, 1]$
$Cosechx$	$\mathbb{R} - \{0\}$	$\mathbb{R} - \{0\}$

Hyperbolic Identities:

(i) $Cosh^2 x - Sinh^2 x = 1 \; \forall x \in \mathbb{R}$

(ii) $Sech^2 x + Tanh^2 x = 1 \; \forall x \in \mathbb{R}$

(iii) $Coth^2 x - Cosech^2 x = 1 \; \forall x \in \mathbb{R} - \{0\}$

(iv) $Sechx . Coshx = 1 \; \forall x \in \mathbb{R}$

(v) $Cosecx . Sinhx = 1 \; \forall x \in \mathbb{R} - \{0\}$

(vi) $Cothx . Tanhx = 1 \forall x \in \mathbb{R} - \{0\}$

Hyperbolic Functions for negative values:

(i) $Sinh(-x) = -Sinhx$

(ii) $Cosh(-x) = Coshx$

(iii) $Tanh(-x) = -Tanhx$

(iv) $Coth(-x) = -Cothx$

(v) $Sech(-x) = Sechx$

(vi) $Cosech(-x) = -Cosechx$

Properties of Hyperbolic Functions:

(i) $Sinh(x \pm y) = Sinhx\, Coshy \pm Coshx\, Sinhy$

(ii) $Cosh(x \pm y) = Coshx\, Coshy \pm Sinhx\, Sinhy$

(iii) $Tanh(x \pm y) = \dfrac{Tanhx \pm Tanhy}{1 \pm Tanhx\, Tanhy}$

(iv) $Coth(x \pm y) = \dfrac{Cothx\, Cothy \pm 1}{Cothy \pm Cothx}$

(v) $Sinh2x = 2Sinhx\, Coshx = \dfrac{2Tanhx}{1 - Tanh^2 x}$

(vi) $Cosh2x = Cosh^2 x + Sinh^2 x = \dfrac{1 + Tanh^2 x}{1 - Tanh^2 x}$

(vii) $Tanh2x = \dfrac{2Tanhx}{1 + Tanh^2 x}$

(viii) $Sinh3x = 3Sinhx + 4Sinh^3 x$

(ix) $Cosh3x = 4Cosh^3 x - 3Coshx$

(x) $Tanh3x = \dfrac{3Tanhx + Tanh^3 x}{1 + 3Tanh^2 x}$

(xi) $Sinh(x+y)Sinh(x-y) = Sinh^2 x - Sinh^2 y$

(xii) $Cosh(x+y)Cosh(x-y) = Cosh^2 x + Sinh^2 y$

(xiii) $(Coshx + Sinhx)^n = Coshnx + Sinhx = e^{nx}$

(xiv) $(Coshx - Sinhx)^n = Coshnx - Sinhx = e^{-nx}$

Inverse Hyperbolic Functions:

Inverse Hyperbolic Function	Domain	Range
$Sinh^{-1}x$	$\mathbb{R}$	$\mathbb{R}$
$Cosh^{-1}x$	$[1, \infty)$	$[0, \infty)$
$Tanh^{-1}x$	$(-1, 1)$	$\mathbb{R}$
$Coth^{-1}x$	$\mathbb{R} - [-1, 1]$	$\mathbb{R} - \{0\}$
$Sech^{-1}x$	$(0, 1]$	$[0, \infty)$
$Cosech^{-1}x$	$\mathbb{R} - \{0\}$	$\mathbb{R} - \{0\}$

(i) $Sinh^{-1}(x) = \log_e(x + \sqrt{x^2+1}) \quad \forall x \in \mathbb{R}$

(ii) $Cosh^{-1}(x) = \log_e(x + \sqrt{x^2-1}) \quad \forall x \geq 1$

(iii) $Tanh^{-1}x = \frac{1}{2}\log\left(\frac{1+x}{1-x}\right) \quad \forall x \in (-1,1)$

(iv) $Coth^{-1}x = \frac{1}{2}\log\left(\frac{x+1}{x-1}\right) \quad \forall x \in \mathbb{R} - [-1,1]$

Euler's Formulae:

(i) $\quad Coshx + Sinhx = e^{ix}$

(ii) $\quad Coshx - Sinhx = e^{-ix}$

Properties of Euler's formula:

(i) $Sinh(ix) = iSinx$

(ii) $Cosh(ix) = Cosx$

(iii) $Tanh(ix) = iTanx$

(iv) $Coth(ix) = -iCotx$

(v) $Sech(ix) = Secx$

(vi) $Cosech(ix) = -iCoseccx$

27. COMPLEX NUMBERS AND DEMOIVRE'S THEOREM

Complex Number: A number is in the form of $z = a + ib$ where a, b are real numbers and i (*iota*) is a non-real number given by $i^2 = -1$ is called as a complex number.

Here a, b are called real par and imaginary part of z respectively. They are denoted by $Re(z)$ and $Im(z)$ respectively.

Integral powers of iota:

$i = \sqrt{-1}$ $i^2 = -1$ $i^3 = -i$ $i^4 = 1$

$i^{4n} = 1$ $i^{4n+1} = i$ $i^{4n+2} = -1$ $i^{4n+3} = -i$ $i^{4n+4} = 1$

Conjugate of a complex number: If $z = a + ib$ is a complex number then $\bar{z} = a - ib$ is denoted as the conjugate of z

Additive inverse of a complex number: If $z = a + ib$ is a complex number then $-z = -a - ib$ is denoted as the additive inverse of z

Multiplicative inverse of a complex number: If $z = a + ib$ is a complex number then $\frac{1}{z} = \frac{1}{a+ib} = \frac{a-ib}{a^2+b^2}$ is denoted as the multiplicative inverse of z

Algebric operations on complex numbers:

If $z_1 = a_1 + ib_1$ and $z_2 = a_2 + ib_2$ are two complex numbers. Then

(i) $z_1 + z_2 = (a_1 + a_2) + i(b_1 + b_2)$

(ii) $z_1 - z_2 = (a_1 - a_2) + i(b_1 - b_2)$

(iii) $z_1 z_2 = (a_1 a_2 - b_1 b_2) + i(a_1 b_2 + a_2 b_1)$

(iv) $\frac{z_1}{z_2} = \frac{a_1 + ib_1}{a_2 + ib_2} = \frac{(a_1 + ib_1)(a_2 - ib_2)}{a_2^2 + b_2^2}$

(v) $z_1 = z_2$ iff $a_1 = a_2$ and $b_1 = b_2$

Properties of conjugate of a complex number:

If z is a complex number then

(i) $\overline{(\bar{z})} = z$ (ii) $z + \bar{z} = 2Re(z)$ (iii) $z - \bar{z} = 2Im(z)$

If z_1 and z_2 are complex numbers. Then

(i) $\overline{z_1 + z_2} = \bar{z_1} + \bar{z_2}$ (ii) $\overline{z_1 - z_2} = \bar{z_1} - \bar{z_2}$

(iii) $\overline{z_1 . z_2} = \bar{z_1} . \bar{z_2}$ (iv) $\overline{\left(\frac{z_1}{z_2}\right)} = \frac{\bar{z_1}}{\bar{z_2}}$

(v) $z_1 = z_2$ iff $\bar{z_1} = \bar{z_2}$

Square root of a complex number: If $z = a + ib$ is a complex number then the square root of z is given by $\sqrt{a + ib} = \pm \left[\frac{\sqrt{\sqrt{a^2+b^2}+a}}{2} + i \frac{\sqrt{\sqrt{a^2+b^2}-a}}{2} \right]$

The square root of $a - ib$ is $\sqrt{a - ib} = \pm \left[\frac{\sqrt{\sqrt{a^2+b^2}+a}}{2} - i \frac{\sqrt{\sqrt{a^2+b^2}-a}}{2} \right]$

Modulus of a complex number: If $z = a + ib$ is a complex number then the modulus of z is denoted by $|z| = a^2 + b^2$

Properties of modulus:

If $z = a + ib$ is a complex number then

    (i) $|z| = 0$ iff z=0

    (ii) $|z| = |-z|$

    (iii) $|z| = |\bar{z}|$

    (iv) $z\bar{z} = |z|^2$

    (v) $|z^n| = |z|^n$

If z_1 and z_2 are complex numbers then

    (i) $|z_1 z_2| = |z_1||z_2|$

    (ii) $\left|\frac{z_1}{z_2}\right| = \frac{|z_1|}{|z_2|}$

    (iii) $|z_1 \pm z_2| \leq |z_1| + |z_2|$

    (iv) $|z_1 \pm z_2| \geq ||z_1| - |z_2||$

    (v) $|z_1 \pm z_2|^2 = |z_1|^2 + |z_2|^2 \pm 2\, Re(z_1 \bar{z_2})$

    (vi) $|z_1| - |z_2| \leq |z_1 + z_2| \leq |z_1| + |z_2|$

Argument (or) Amplitude of a complex number: If $z = a + ib$ is a complex number then the argument or amplitude is denoted as $Arg(z)$ or $Amp(z) = Tan^{-1}\left(\frac{b}{a}\right)$

Argument in various quadrants:

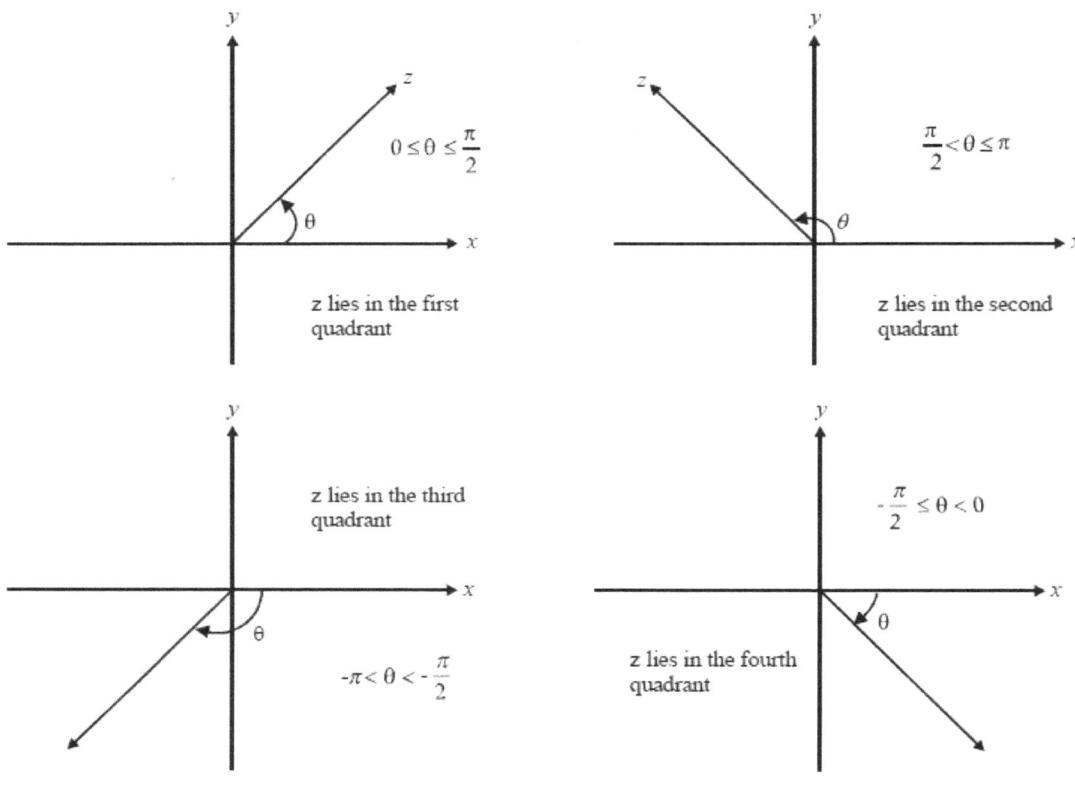

Properties of Argument:

If z, z_1, z_2 are complex numbers, then

(i) $Arg(\bar{z}) = -Arg(z)$

(ii) $Arg(z) + Arg(\bar{z}) = \begin{cases} 0 \text{ if } \theta \neq \pi \\ 2\pi \text{ if } \theta = \pi \end{cases}$

(iii) $Arg(z_1 z_2) = Arg(z_1) + Arg(z_2) + 2k\pi \text{ for } k = \{-1, 0, 1\}$

(iv) $Arg\left(\dfrac{z_1}{z_2}\right) = Arg(z_1) - Arg(z_2) + 2k\pi \text{ for } k = \{-1, 0, 1\}$

(v) $Arg(z_1 \bar{z}_2) = Arg(z_1) - Arg(z_2)$

(vi) $Arg(z^n) = n \, Arg(z)$

Polar Form (or) Principal Modulus-Amplitude form:

If $z = a + ib$ is a complex number then the polar form of z is represented as

$z = r(Cos\theta + i\, Sin\theta) = rCis\theta$ where $r = \sqrt{a^2 + b^2}$ is the modulus of z

and $\theta = Tan^{-1}\left(\dfrac{b}{a}\right)$ is the amplitude of z.

General Modulus-Amplitude form:

If $z = a + ib$ is a complex number then the general modulus-amplitude form of z is represented as $z = rCis(2k\pi + \theta)$ where $k \in I$

Properties:

If z_1, z_2 are complex numbers then

(i) $|z_1 + z_2|^2 = r_1^2 + r_2^2 + 2r_1 r_2 Cos(\theta_1 - \theta_2)$

(ii) $|z_1 - z_2|^2 = r_1^2 + r_2^2 - 2r_1 r_2 Cos(\theta_1 - \theta_2)$

(iii) $|z_1 + z_2|^2 + |z_1 - z_2|^2 = 2(r_1^2 + r_2^2)$

(iv) $|z_1 + z_2|^2 - |z_1 - z_2|^2 = 4r_1 r_2 Cos(\theta_1 - \theta_2)$

Euler's Form:

The complex number $z = r(Cos\theta + i\, Sin\theta) = rCis\theta$ can be denoted in Euler's form is $z = re^{i\theta}$

Properties:

(i) $e^{i\theta} = Cos\theta + i\, Sin\theta$ (ii) $e^{-i\theta} = Cos\theta - i\, Sin\theta$ (iii) $\dfrac{1}{Cis\theta} = Cis(-\theta)$

(iv) $Cis\theta_1 . Cis\theta_2 = Cis(\theta_1 + \theta_2)$ (v) $\dfrac{Cis\theta_1}{Cis\theta_2} = Cis(\theta_1 - \theta_2)$

(vi) $Cos\theta = \dfrac{e^{i\theta} + e^{-i\theta}}{2} = Coshi\theta$ (vii) $i\, Sin\theta = \dfrac{e^{i\theta} - e^{-i\theta}}{2} = Sinhi\theta$

Argand diagram: The diagram formed by the given complex numbers in argand plane is called as an argand diagram

Geometrical applications to complex numbers:

(i) The distance between the two points $A(z_1)$ and $B(z_2)$ is $AB = |z_2 - z_1|$

i.e. The points A and B represented by the complex numbers $z_1 = a_1 + ib_1$ and $z_2 = a_2 + ib_2$ respectively then $AB = |z_2 - z_1| = \sqrt{(x_2 - x_1)^2 + (y_2 - y_1)^2}$

(ii) The point $P(z)$ divides the line joining the line segment joining the points $A(z_1)$ and $B(z_2)$ in the ratio $m:n$ internally then $z = \dfrac{mz_2 + nz_1}{m+n}$

(iii) The point $Q(z)$ divides the line joining the line segment joining the points $A(z_1)$ and $B(z_2)$ in the ratio $m:n$ externally then $z = \dfrac{mz_2 - nz_1}{m-n}$

(iv) The midpoint $M(z)$ of the line segment joining the points $A(z_1)$ and $B(z_2)$ then $z = \dfrac{z_1+z_2}{2}$

(v) Collinear Points: The points represented by the complex numbers z_1, z_2, z_3 are collinear when $\begin{vmatrix} z_1 & \bar{z}_1 & 1 \\ z_2 & \bar{z}_2 & 1 \\ z_3 & \bar{z}_3 & 1 \end{vmatrix} = 0$ or z_1, z_2, z_3 are in A.P.

(vi) Triangle: Let $A(z_1)$, $B(z_2)$ and $C(z_3)$ are the vertices of a triangle then

Centroid = $\dfrac{z_1+z_2+z_3}{3}$ Incentre = $\dfrac{az_1+bz_2+cz_3}{a+b+c}$ where $a = |z_2 - z_3|$; $b = |z_3 - z_1|$; $c = |z_1 - z_2|$

Circumcentre = $\dfrac{\sum z_1 \, Sin2A}{\sum Sin2A}$

Orthocentre = $\dfrac{\sum z_1 \, TanA}{\sum TanA}$

(vii) Area of Triangle:

The area of the triangle whose vertices are $z, iz, z + iz$ is $\dfrac{1}{2}|z|^2$ sq. units

The area of the triangle whose vertices are $z, iz, z - iz$ is $\dfrac{3}{2}|z|^2$ sq. units

The area of the triangle whose vertices are $z, \omega z, z + \omega z$ is $\dfrac{\sqrt{3}}{4}|z|^2$ sq. units

(viii) Quadrilateral: Let $A(z_1)$, $B(z_2)$, $C(z_3)$ and $D(z_4)$ are the vertices of a quadrilateral then

Parallelogram $\Leftrightarrow z_1 + z_3 = z_2 + z_4$

Rectangle $\Leftrightarrow z_1 + z_3 = z_2 + z_4$ and $|z_1 - z_3| = |z_2 - z_4|$

Rhombus $\Leftrightarrow z_1 + z_3 = z_2 + z_4$ and

$|z_1 - z_2| = |z_2 - z_3| = |z_3 - z_4| = |z_4 - z_1|$

Square $\Leftrightarrow z_1 + z_3 = z_2 + z_4$ and $|z_1 - z_3| = |z_2 - z_4|$

$|z_1 - z_2| = |z_2 - z_3| = |z_3 - z_4| = |z_4 - z_1|$

Locus of given complex numbers:

(i) Straight Line:

(a) The equation of straight line passes through the points $A(z_1)$, $B(z_2)$ is $\begin{vmatrix} z & \bar{z} & 1 \\ z_1 & \bar{z}_1 & 1 \\ z_2 & \bar{z}_2 & 1 \end{vmatrix} = 0$

(b) The general equation of straight line is $\bar{a}z + a\bar{z} + b = 0$

(ii) Circle:

   (a) The equation of circle having the centre z_0 and the radius r is
$|z - z_0| = r$ (or) $z\bar{z} - z_0\bar{z} - \bar{z_0}z + z_0\bar{z_0} = r^2$

   (b) The equation of circle whose end points of a diameter are z_1, z_2 is
$|z - z_1|^2 + |z - z_2|^2 = |z_1 - z_2|^2$

   (c) The general equation of the circle is $z\bar{z} + \bar{a}z + a\bar{z} + b = 0$

(iii) Miscellaneous:

   (a) The equation $\left|\frac{z-z_1}{z-z_2}\right| = k$ $(k \neq 1)$ represents a circle

   (b) The equation $\left|\frac{z-z_1}{z-z_2}\right| = k$ $(k = 1)$ represents a perpendicular bisector of line joining z_1, z_2

   (c) The equation $Arg\left(\frac{z-z_1}{z-z_2}\right) = \pm\frac{\pi}{2}$ represents a circle whose end points of a diameter are z_1, z_2

   (d) The equation $Arg\left(\frac{z-z_1}{z-z_2}\right) = 0 \ or \ \pi$ represents a straight line joining the points z_1, z_2

(iv) Conics:

   Circle: $|z - z_1|^2 + |z - z_2|^2 = k \in \mathbb{R}$ where $k \geq \frac{1}{2}|z_1 - z_2|$ represents a circle

   Ellipse: $|z - z_1| + |z - z_2| = k$ represents

     (a) An ellipse when $k > |z_1 - z_2|$

     (b) An empty set when $k < |z_1 - z_2|$

     (c) A line segment when $k = |z_1 - z_2|$

   Hyperbola: $||z - z_1| - |z - z_2|| = k$ represents

     (a) A hyperbola when $k < |z_1 - z_2|$

     (b) An empty set when $k > |z_1 - z_2|$

     (c) Two rays when $k = |z_1 - z_2|$

DeMoivre's Theorem

DeMoivre's Theorem: If n is a positive integer

   (i) $(Cos\theta + i\, Sin\theta)^n = Cosn\theta + i\, Sinn\theta = Cis(n\theta)$

   (ii) $(Cos\theta - i\, Sin\theta)^n = Cosn\theta - i\, Sinn\theta = Cis(-n\theta)$

   (iii) $(Cos\theta + i\, Sin\theta)^{-n} = Cosn\theta - i\, Sinn\theta = Cis(-n\theta)$

   (iv) $(Cos\theta + i\, Sin\theta)^{1/n} = Cos\left(\frac{2k\pi+\theta}{n}\right) + i\, Sin\left(\frac{2k\pi+\theta}{n}\right) = Cis\left(\frac{2k\pi+\theta}{n}\right)$

   where $k = 0,1,2,\ldots\ldots(n-1)$

   (v) $(Cos\theta + i\, Sin\theta)^{p/q} = Cos\left(\frac{2k\pi+p\theta}{q}\right) + i\, Sin\left(\frac{2k\pi+p\theta}{q}\right) = Cis\left(\frac{2k\pi+p\theta}{q}\right)$

   where $k = 0,1,2,\ldots\ldots(q-1)$

Properties of DeMoivre's Theorem:

(i) If $x = Cis\theta$ then

(a) $\frac{1}{x} = Cis(-\theta)$ (b) $x + \frac{1}{x} = 2Cos\theta$ (c) $x - \frac{1}{x} = 2i Sin\theta$

(d) $x^n + \frac{1}{x^n} = 2 Cosn\theta$ (e) $x^n - \frac{1}{x^n} = 2i Sinn\theta$

(ii) If $Cos\alpha + Cos\beta + Cos\gamma = Sin\alpha + Sin\beta + Sin\gamma = 0$ then

(a) $Cos^2\alpha + Cos^2\beta + Cos^2\gamma = Sin^2\alpha + Sin^2\beta + Sin^2\gamma = \frac{3}{2}$

(b) $Cos3\alpha + Cos3\beta + Cos3\gamma = 3Cos(\alpha + \beta + \gamma)$

$Sin3\alpha + Sin3\beta + Sin3\gamma = 3Sin(\alpha + \beta + \gamma)$

(c) $Cos2^n\alpha + Cos2^n\beta + Cos2^n\gamma = Sin2^n\alpha + Sin2^n\beta + Sin2^n\gamma = 0$

(d) $Cos(\alpha - \beta) + Cos(\beta - \gamma) + Cos(\gamma - \alpha)$
$= Sin(\alpha - \beta) + Sin(\beta - \gamma) + Sin(\gamma - \alpha) = 0$

n^{th} roots of Unity:

The nth roots of unity are given by $(1)^{1/n} = Cis\left(\frac{2k\pi}{n}\right)$ where $k = 0,1,2,\ldots\ldots(n-1)$

Square roots of Unity:

The square roots of unity are given by $(1)^{1/2} = Cis(k\pi)$ where $k = 0,1$ i.e. $(1)^{1/2} = 1, -1$

Cube roots of Unity:

The cube roots of unity are given by $(1)^{1/3} = Cis\left(\frac{2k\pi}{3}\right)$ where $k = 0,1,2$

i.e. $(1)^{1/3} = 1, \omega, \omega^2$ where $\omega = \frac{-1+\sqrt{3}i}{2}$ and $\omega^2 = \frac{-1-\sqrt{3}i}{2}$

Here, $1 + \omega + \omega^2 = 0$ and $\omega^3 = 1$

Properties:

(i) The roots of $x^2 + x + 1 = 0$ are ω, ω^2

(ii) The roots of $x^2 - x + 1 = 0$ are $-\omega, -\omega^2$

(iii) The roots of $x^2 + 2x + 4 = 0$ are $2\omega, 2\omega^2$

(iv) The roots of $x^2 - 2x + 4 = 0$ are $-2\omega, -2\omega^2$

4th roots of Unity:

The fourth roots of unity are given by $(1)^{1/4} = Cis\left(\frac{k\pi}{2}\right)$ where $k = 0,1,2,3$

i.e. $(1)^{1/4} = 1, -1, i, -i$

2-DIMENSIONAL GEOMETRY

28. 2D CO-ORDINATES

1. Distance between two points:

(i) The distance between the two points $A(x_1, y_1)$ and $B(x_2, y_2)$ is

$$AB = \sqrt{(x_2 - x_1)^2 + (y_2 - y_1)^2}$$

(ii) The distance between the two points $O(0,0)$ and $A(x_1, y_1)$ is

$$OA = \sqrt{x_1^2 + y_1^2}$$

2. Slope:
The slope of the line segment joining the points $A(x_1, y_1)$ and $B(x_2, y_2)$ is

$$m = \frac{y_2 - y_1}{x_2 - x_1}$$

3. Collinear Points:
Three are more points are said to be collinear when they lie on a same line.

To show $A(x_1, y_1), B(x_2, y_2)$ and $C(x_3, y_3)$ are collinear we have to show any one of the following

(i) Slope of AB= Slope of BC= Slope of CA

(ii) Area of triangle ABC=0

(iii) The sum of the lengths of any two sides is equal to the length of the third side.

i.e. $AB + BC = CA$ or $BC + CA = AB$ or $CA + AB = BC$

4. Section Formula:

(i) If a point P divides the line joining the points $A(x_1, y_1), B(x_2, y_2)$ in the ratio $m:n$ internally then

$$P = \left(\frac{mx_2 + nx_1}{m + n}, \frac{my_2 + ny_1}{m + n}\right)$$

(ii) If a point Q divides the line joining the points $A(x_1, y_1), B(x_2, y_2)$ in the ratio $m:n$ externally then

$$Q = \left(\frac{mx_2 - nx_1}{m - n}, \frac{my_2 - ny_1}{m - n}\right)$$

5. Midpoint:
The midpoint of the line segment joining the points $A(x_1, y_1), B(x_2, y_2)$ is $\left(\frac{x_1+x_2}{2}, \frac{y_1+y_2}{2}\right)$

6. Points of Trisection:
The points which divides the line segment joining two points in the ratio $1:2$ or $2:1$ are called as the points of trisection.

7. Harmonic Conjugate:
If the points P and Q divides the line segment joining tow points in the same ratio internally and externally respectively then one of P and Q is called as harmonic conjugate to the other.

8. Triangles: Let $A(x_1, y_1)$, $B(x_2, y_2)$ and $C(x_3, y_3)$ be the vertices of a triangle ABC then

(i) $\triangle ABC$ is a right angled triangle when the square of one side is equal to the sum of the squares of other two sides

(ii) $\triangle ABC$ is an acute angled triangle when the square of one side is less than the sum of the squares of other two sides

(iii) $\triangle ABC$ is an obtuse angled triangled when the square of one side is greater than the sum of the squares of other two sides

(iv) $\triangle ABC$ is a scalene triangle when the lengths of all sides are different

(v) $\triangle ABC$ is an isosceles triangle when the lengths of two sides are equal

(vi) $\triangle ABC$ is an equilateral triangle when the lengths of three sides are equal

9. Quadrilaterals: Let $A(x_1, y_1)$, $B(x_2, y_2)$, $C(x_3, y_3)$ and $D(x_4, y_4)$ be the vertices of a quadrilateral ABCD. Then

(i) ABCD is a Parallelogram when Midpoint of AC=Midpoint of BD

(ii) ABCD is a Rhombus when

Midpoint of AC=Midpoint of BD, AB=BC and $AC \neq BD$

(iii) ABCD is a Rectangle when

Midpoint of AC=Midpoint of BD, $AB \neq BC$ and $AC = BD$

(iv) ABCD is a Square when

Midpoint of AC=Midpoint of BD, $AB = BC$ and $AC = BD$

10. Area of a triangle:

(i) The area of a triangle whose vertices are $A(x_1, y_1)$, $B(x_2, y_2)$ and $C(x_3, y_3)$ is $\frac{1}{2}|x_1(y_2 - y_3) + x_2(y_3 - y_1) + x_3(y_1 - y_2)|$ or

$$\frac{1}{2}\begin{vmatrix} x_1 & y_1 & 1 \\ x_2 & y_2 & 1 \\ x_3 & y_3 & 1 \end{vmatrix} \text{ or } \frac{1}{2}\begin{vmatrix} x_1 - x_2 & y_1 - y_2 \\ x_1 - x_3 & y_1 - y_3 \end{vmatrix} \text{ sq. units.}$$

(ii) The area of a triangle whose vertices are $O(0,0), A(x_1, y_1)$, $B(x_2, y_2)$ is $\frac{1}{2}|x_1 y_2 - x_2 y_1|$ sq. units.

11. Area of a quadrilateral:

The area of a quadrilateral whose vertices are $A(x_1, y_1)$, $B(x_2, y_2)$, $C(x_3, y_3)$ and $D(x_4, y_4)$ is $\frac{1}{2}\begin{vmatrix} x_1 - x_3 & y_1 - y_3 \\ x_2 - x_4 & y_2 - y_4 \end{vmatrix}$ sq. units.

12. Concurrent lines and Point of concurrency: Two are more lines are said to be concurrent when they passes through same point. That point is called point of concurrency.

13. Median: The line segment joining the vertex and midpoint of the opposite side is called as a median.

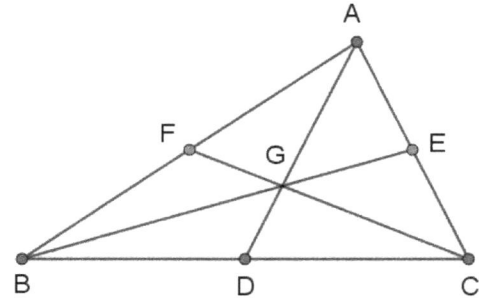

Lengths of the medians: Let AD, BE, CF be the medians drawn from A, B, C respectively

(i) $AD = \frac{1}{2}\sqrt{2b^2 + 2c^2 - a^2}$

(ii) $BE = \frac{1}{2}\sqrt{2c^2 + 2a^2 - b^2}$

(iii) $CF = \frac{1}{2}\sqrt{2a^2 + 2b^2 - c^2}$ where $a = BC\ ; b = CA\ ; c = AB$

14. Centroid: The point of concurrency of three medians of a triangle is called as a centroid. It is denoted by G.

G divides each median in the ratio 2:1 internally.

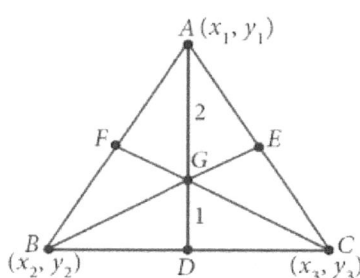

The centroid of the triangle formed by the points $A(x_1, y_1)$, $B(x_2, y_2)$ and $C(x_3, y_3)$ is

$G = \left(\dfrac{x_1 + x_2 + x_3}{3}, \dfrac{y_1 + y_2 + y_3}{3}\right)$

15. Altitude: The perpendicular drawn from a vertex to the opposite side is called as an altitude.

16. Orthocentre: The point of concurrency of three altitudes of a triangle is called as a orthocentre. It is denoted by O.

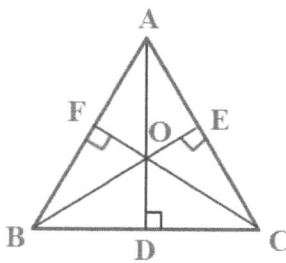

17. Perpendicular Bisector: The line passes through midpoint of a side and perpendicular to the side of a triangle is called as a perpendicular bisector.

18. Circumcentre: The point of concurrency of three perpendicular bisectors of a triangle is called as a circumcentre. It is denoted by S.

$SA = SB = SC$

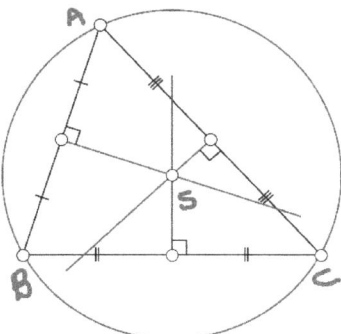

19. Circumcircle: The circle which passes through the three vertices of a triangle is called as a circumcircle.

The radius of a circumcircle of a triangle ABC is $SA = SB = SC = R$

20. Internal Angular Bisector: The line which bisects the internal angle of a vertex is called as an internal angular bisector.

The internal angular bisector of A of triangle ABC divides the Side BC in the ratio AB:AC internally

21. Incentre: The point of concurrency of three internal angular bisectors of a triangle is called as a incentre. It is denoted by I.

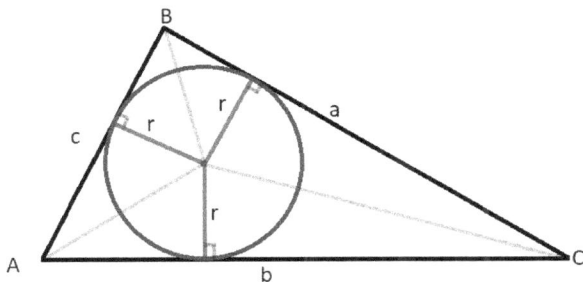

The incentre of the triangle formed by the points $A(x_1, y_1)$, $B(x_2, y_2)$ and $C(x_3, y_3)$ is

$$I = \left(\frac{ax_1 + bx_2 + cx_3}{a + b + c}, \frac{ay_1 + by_2 + cy_3}{a + b + c}\right)$$

where $a = BC$; $b = CA$; $c = AB$

22. Incircle: The circle which touches all the sides is called as incircle.

Inradius is denoted by r.

23. External angular bisector: The line which bisects the external angle of a vertex is called as an external angular bisector.

24. Excentre:
The point of concurrency of internal angular bisector of an angle and external angular bisectors of other two angles of a triangle is called as a excentre.

The excentres opposite to the vertices A, B, C are denoted by I_A, I_B, I_C respectively.

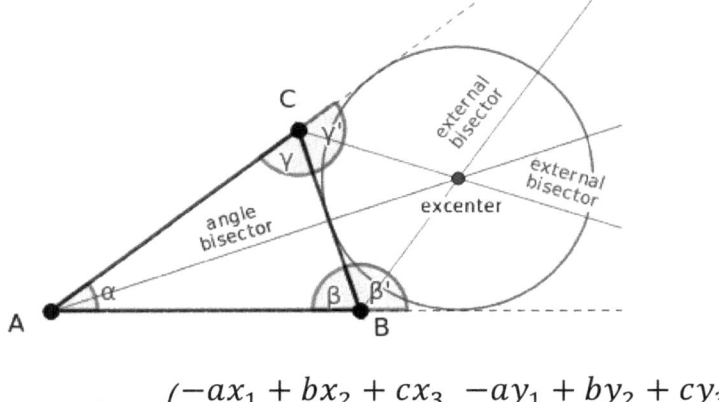
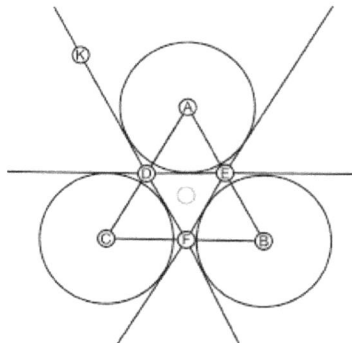

$$I_A = \left(\frac{-ax_1 + bx_2 + cx_3}{-a + b + c}, \frac{-ay_1 + by_2 + cy_3}{-a + b + c}\right)$$

$$I_B = \left(\frac{ax_1 - bx_2 + cx_3}{a - b + c}, \frac{ay_1 - by_2 + cy_3}{a - b + c}\right)$$

$$I_C = \left(\frac{ax_1 + bx_2 - cx_3}{a + b - c}, \frac{ay_1 + by_2 - cy_3}{a + b - c}\right)$$

25. Relation between G, O, S:

In any triangle G, O, S are collinear and G divides the line segment joining O and S in the ratio 2:1 internally.

i.e. $OG:GS = 2:1$ and $3G = 2S + O$

26. Miscellaneous:

(i) In an equilateral triangle, G, O, S, I all are coincide

(ii) In a right angled triangle, orthocentre is the vertex at right angle and the circumcentre is the midpoint of hypotenuse.

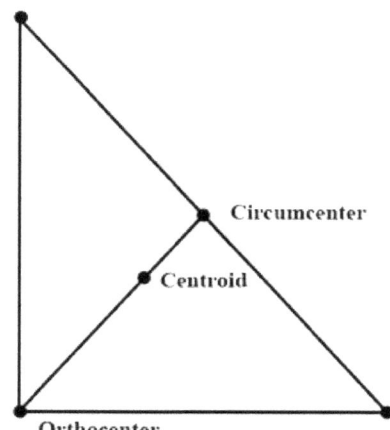

27. Nine point circle:
Let ABC be a triangle. Let D, E, F be the feet of the altitudes and X, Y, Z be the midpoints of the sides. Let P, Q, R be the midpoints of AO, BO, CO where O is the orthocentre of the triangle. Then the circle passes through D, E, F, X,

Y, Z, P, Q, R is called as a nine point circle. The centre of nine point circle is denoted by N.

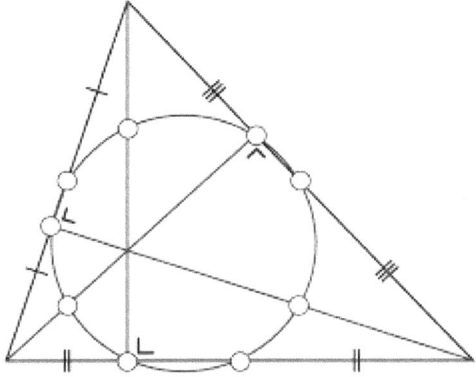

O, N, G, S are collinear points with $ON:NG:GS = 3:1:2$

29. LOCUS

Locus: The path which joins the points which obey the given geometrical condition is called as a locus.

Eg: The locus of the points which are at a constant distance from a fixed point is a circle.

Various types of Locus:

(i) The locus of a point which is equidistant from two fixed points A and B is the perpendicular bisector of the line segment AB.

(ii) The locus of the points which are at a constant distance from a fixed point is a circle.

(iii) The locus of the third vertex of a right-angled triangle is the circle whose end points of diameter are the ends of hypotenuse of the triangle.

(iv) Let A and B be two fixed points. Let P be a point moves such that $\frac{PA}{PB} = k$. Then the locus of P is

   (a) a straight line when $k = 1$

   (b) a circle when $k > 0$ and $k \neq 1$

   (c) an empty set when $k < 0$

(v) Let A, B, C be the fixed points. Then the locus of P such that
   $PA^2 + PB^2 = k.PC^2$ is

   (a) a straight line when $k = 2$

   (b) a circle when $k > 0$ and $k \neq 2$

   (c) an empty set when $k < 0$

(vi) Let A and B be two fixed points. Let P be a point moves such that $PA + PB = k$. Then the locus of P is

   (a) an ellipse when $AB < k$

   (b) a line segment when $AB = k$

   (c) does not exist when $AB > k$

(vii) Let A and B be two fixed points. Let P be a point moves such that
   $|PA - PB| = k$. Then the locus of P is

   (a) a hyperbola when $AB > k$

   (b) a line segment when $AB = k$

   (c) does not exist when $AB < k$

30. CHANGE OF AXES

Translation of axes: The process that shifting the origin to some another point without changing the direction of axes is called as the translation of axes.

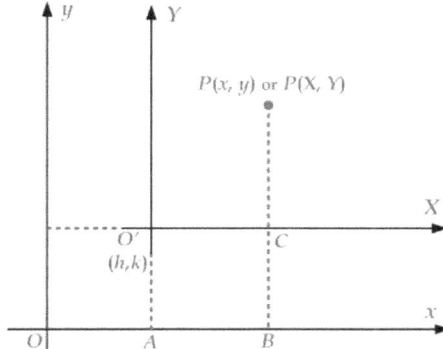

Let we change the origin to a point (h, k) through translation and the original point (x, y) transformed to the point (X, Y) then $x = X + h$ and $y = Y + k$

Rotation of axes: The process that rotating the angle between the co-ordinate axes without changing the position of the origin is called as the rotation of axes.

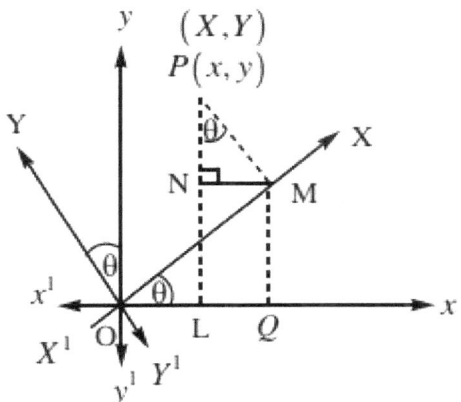

Let we rotate the axes through an angle θ in anticlockwise direction. Then the transformed point (X, Y) of (x, y) is given as follows.

	X	Y
x	$Cos\theta$	$-Sin\theta$
y	$Sin\theta$	$Cos\theta$

From the above table, we conclude that

$x = X\,Cos\theta - Y\,Sin\theta \;\; ; y = X\,Sin\theta + Y\,Cos\theta$

$X = x\,Cos\theta + y\,Sin\theta \;\; ; Y = -x\,Sin\theta + y\,Cos\theta$

Applications of change of axes:

(i) To remove the first degree terms from the equation

$S \equiv ax^2 + 2hxy + by^2 + 2gx + 2fy + c = 0$, the origin is shifted to the point $\left(\dfrac{hf - bg}{ab - h^2}, \dfrac{hg - af}{ab - h^2}\right)$ where $ab - h^2 \neq 0$

It will be obtain easily to solve the equations $\dfrac{\partial s}{\partial x} = 0$ and $\dfrac{\partial s}{\partial y} = 0$

(ii) To remove the first degree terms from the equation

$ax^2 + by^2 + 2gx + 2fy + c = 0$, the origin is shifted to the point $\left(\dfrac{-g}{a}, \dfrac{-f}{b}\right)$

(iii) To remove the first degree terms from the equation

$2hxy + 2gx + 2fy + c = 0$, the origin is shifted to the point

$\left(\dfrac{-f}{h}, \dfrac{-g}{h}\right)$

(iv) To remove the xy term from the equation

$S \equiv ax^2 + 2hxy + by^2 + 2gx + 2fy + c = 0$, the angle of rotation of axes is $\theta = \dfrac{1}{2}Tan^{-1}\left(\dfrac{2h}{a-b}\right)$ if $a \neq b$ and $\theta = (2n+1)\dfrac{\pi}{4} \; \forall n \in Z \; if \; a = b$

31. STRAIGHT LINES

Inclination: The angle θ $(0 \leq \theta < \pi)$ made by a line with positive X-axis in anticlockwise direction (positive direction) is called as an inclination.

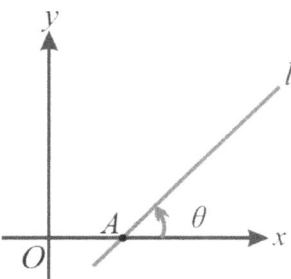

Slope (or) Gradient: If θ is the inclination made by a non vertical line then $Tan\theta$ is called as a slope or gradient of the line. It is denoted by m.

$m = Tan\theta$

Slope of the line passes through $A(x_1, y_1), B(x_2, y_2)$:

The slope of a line passes through the points $A(x_1, y_1), B(x_2, y_2)$ is

$m = \dfrac{y_2 - y_1}{x_2 - x_1}$

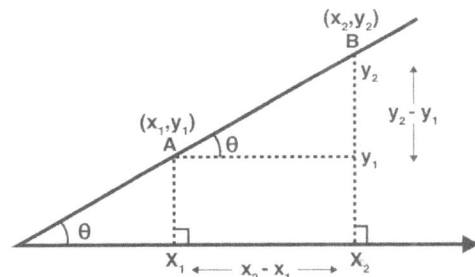

Properties of slopes:

(i) The slope of horizontal line is zero $(\theta = 0^0)$

(ii) The slope of vertical line is undefined $(\theta = 90^0)$

(iii) The slope is positive for acute angle $(0^0 < \theta < 90^0)$

(iv) The slope is negative for obtuse angle $(90^0 < \theta < 180^0)$

Intercepts made by a line with co-ordinate axes:

If a line intersects X and Y axes at A and B then OA is called the X-intercept and OB is called the Y-intercept by the line.

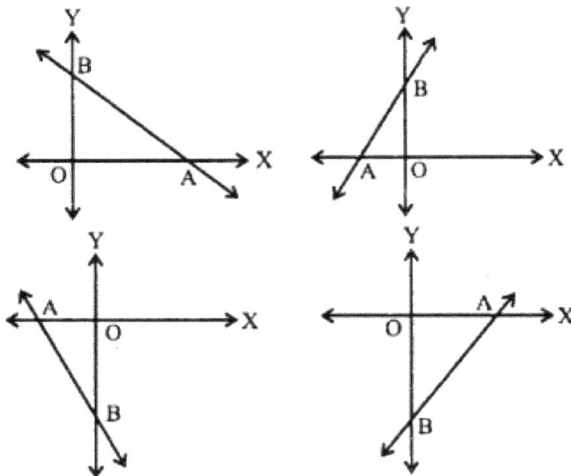

Equations of X and Y axes and their parallel lines:

(i) The equation of X-axis is $y = 0$

(ii) The equation of X-axis is $x = 0$

(iii) The equation of straight line parallel to X-axis is $y = k$ for some scalar k

(iv) The equation of straight line parallel to Y-axis is $x = k$ for some scalar k

Various forms of straight line:

(i) **Slope-Intercept form:** The equation of a straight line having slope m and Y-intercept c is $y = mx + c$

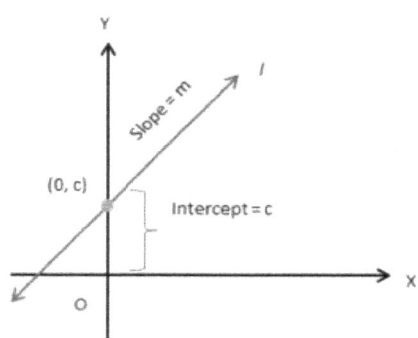

Note: The equation of a straight line which passes through the origin is in the form of $y = mx$

(ii) **Point-Slope form:** The equation of a straight line having slope m and passes through the point (x_1, y_1) is $y - y_1 = m(x - x_1)$

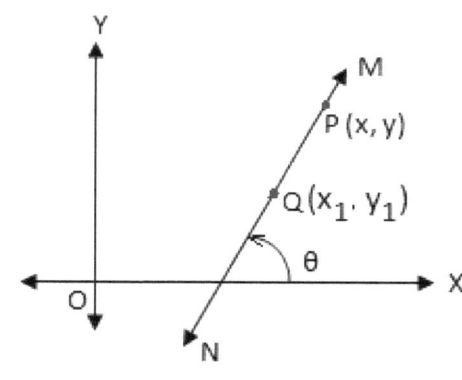

(iii) Two-point form: The equation of a straight line passes through two points

$(x_1, y_1), (x_2, y_2)$ is $y - y_1 = \frac{y_2 - y_1}{x_2 - x_1}(x - x_1)$

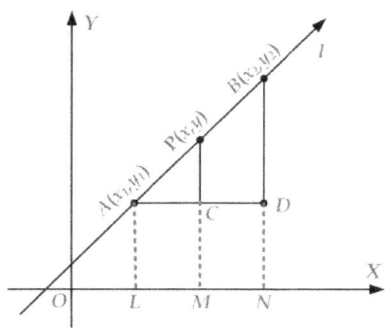

(iv) Intercept form: The equation of a straight line having X and Y intercepts a and b respectively is $\frac{x}{a} + \frac{y}{b} = 1$

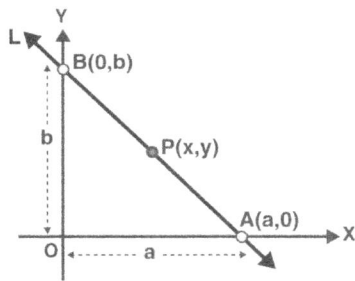

(v) Normal form: The equation of a straight line whose perpendicular distance from origin is p and the perpendicular makes an angle α with the positive X-axis in anticlockwise direction is $x\,Cos\alpha + y\,Sin\alpha = p$

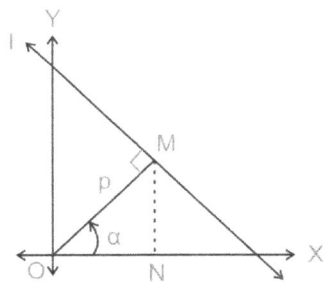

(vi) Parametric form (Symmetric form): The equation of a straight line making an angle θ with the positive X-axis in anticlockwise direction and passes through the point (x_1, y_1) is $\frac{x - x_1}{Cos\theta} = \frac{y - y_1}{Sin\theta} = |r|$

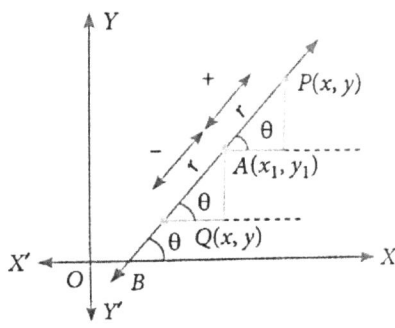

The points which are at a distance of r units from the point (x_1, y_1) are $(x_1 \pm r\,Cos\theta, y_1 \pm r\,Sin\theta)$

(vii) General form: The equation of a straight line in general form is

$ax + by + c = 0\ (a^2 + b^2 \neq 0)$ with a, b, c are real numbers

Here X-intercept is $-c/a$ and Y-intercept is $-c/a$

Slope is $m = -a/b$

Position of a points w.r.to a line: Let $L \equiv ax + by + c = 0$ be a straight line and $A(x_1, y_1), B(x_2, y_2)$ be two points then the ratio that $L = 0$ divides the line segment joining A and B is $-L_{11}: L_{22}$ where $L_{11} = ax_1 + by_1 + c$ and $L_{22} = ax_2 + by_2 + c$

(i) A and B lies same side of the line when L_{11}, L_{22} have same sign

(ii) A and B lies opposite side of the line when L_{11}, L_{22} have opposite sign

Parallel and perpendicular lines: Let $ax + by + c = 0$ be a straight line. Then

(i) The straight line which is parallel to $ax + by + c = 0$ is in the form

$ax + by + k = 0$ for some scalar k

(ii) The straight line which is perpendicular to $ax + by + c = 0$ is in the

form $bx - bay + k = 0$ for some scalar k

(iii) The equation of straight line passes through (x_1, y_1) and parallel to the equation $ax + by + c = 0$ is $a(x - x_1) + b(y - y_1) + c = 0$

(iv) The equation of straight line passes through (x_1, y_1) and perpendicular to the equation $ax + by + c = 0$ is $b(x - x_1) - a(y - y_1) + c = 0$

Foot of the perpendicular: Let $Q(h, k)$ be the foot of the perpendicular drawn from $P(x_1, y_1)$ to the straight line $ax + by + c = 0$ then foot of the perpendicular is given by $\frac{h-x_1}{a} = \frac{k-y_1}{b} = \frac{-(ax_1+by_1+c)}{a^2+b^2}$

Image of a point: Let $Q(h, k)$ be the image of the point $P(x_1, y_1)$ w.r.to. to the straight line $ax + by + c = 0$ then image of the point is given by

$\frac{h-x_1}{a} = \frac{k-y_1}{b} = \frac{-2(ax_1+by_1+c)}{a^2+b^2}$

Distances:

(i) The perpendicular distance from a point (x_1, y_1) to the straight line

$ax + by + c = 0$ is $\frac{|ax_1+by_1+c|}{\sqrt{a^2+b^2}}$

(ii) The perpendicular distance from a point (0,0) to the straight line

$ax + by + c = 0$ is $\frac{|c|}{\sqrt{a^2+b^2}}$

(iii) The distance between the parallel lines $ax + by + c_1 = 0$ and $ax + by + c_2 = 0$ is $\frac{|c_1-c_2|}{\sqrt{a^2+b^2}}$

Point of intersection: The point of intersection of lines

$a_1x + b_1y + c_1 = 0$ and $a_2x + b_2y + c_2 = 0$ is $\left(\frac{b_1c_2-b_2c_1}{a_1b_2-a_2b_1}, \frac{c_1a_2-c_2a_1}{a_1b_2-a_2b_1}\right)$

Concurrent lines and point of concurrency: Three are more lines are said to be concurrent when they passes through the same point. The point is called the point of concurrency.

The three lines $a_1x + b_1y + c_1 = 0$; $a_2x + b_2y + c_2 = 0$ and $a_3x + b_3y + c_3 = 0$ are concurrent then $\begin{vmatrix} a_1 & b_1 & c_1 \\ a_2 & b_2 & c_2 \\ a_3 & b_3 & c_3 \end{vmatrix} = 0$

The point of concurrency is the point of intersection of any two lines.

Family of lines:

(i) Let $L_1 = 0$ and $L_2 = 0$ be two intersecting lines. Then the family of lines consists these lines as members is $L_1 + \lambda L_2 = 0$ where λ is a parameter. These lines passes through the point of intersection of $L_1 = 0$ and $L_2 = 0$

(ii) Let $L_1 = 0$; $L_2 = 0$ and $L_3 = 0$ be intersecting lines. Then the family of lines consists these lines as members is $\lambda_1 L_1 + \lambda_2 L_2 + \lambda_3 L_3 = 0$ where $\lambda_1, \lambda_2, \lambda_3$ are not all zero

Angle between two straight lines:

(i) Let θ be the angle between two straight lines with slopes m_1, m_2. Then θ is given by $Tan\theta = \left|\frac{m_1-m_2}{1+m_1 m_2}\right|$

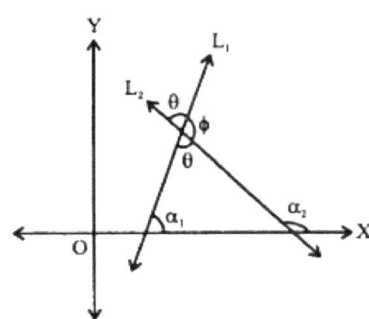

Properties: (a) If $m_1 = m_2$ then the two lines are parallel

(b) If $m_1 m_2 = -1$ then the two lines are perpendicular

(ii) Let θ be the angle between two straight lines $a_1x + b_1y + c_1 = 0$; $a_2x + b_2y + c_2 = 0$ then θ is given by

$$Cos\theta = \frac{|a_1a_2 + b_1b_2|}{\sqrt{a_1^2 + b_1^2}\sqrt{a_2^2 + b_2^2}}$$

Properties: (a) If $a_1 a_2 + b_1 b_2 = 0$ then the lines are perpendicular

(b) If $\frac{a_1}{a_2} = \frac{b_1}{b_2}$ then the lines are parallel

(c) If $\frac{a_1}{a_2} = \frac{b_1}{b_2} = \frac{c_1}{c_2}$ then the lines are coincident

Area of triangle:

(i) The area of a triangle whose vertices are $A(x_1, y_1)$, $B(x_2, y_2)$ and $C(x_3, y_3)$ is $\frac{1}{2}|x_1(y_2 - y_3) + x_2(y_3 - y_1) + x_3(y_1 - y_2)|$ or

$$\frac{1}{2}\begin{vmatrix} x_1 & y_1 & 1 \\ x_2 & y_2 & 1 \\ x_3 & y_3 & 1 \end{vmatrix} \text{ or } \frac{1}{2}\begin{vmatrix} x_1 - x_2 & y_1 - y_2 \\ x_1 - x_3 & y_1 - y_3 \end{vmatrix} \text{ sq. units.}$$

(ii) The area of a triangle whose vertices are $O(0,0), A(x_1, y_1), B(x_2, y_2)$ is $\frac{1}{2}|x_1 y_2 - x_2 y_1|$ sq. units.

(iii) The area of the triangle formed by the straight line $ax + by + c = 0$ with the co-ordinate axes is $\frac{c^2}{2|ab|}$

(iv) The area of the triangle formed by the straight line $\frac{x}{a} + \frac{y}{b} = 1$ with the co-ordinate axes is $\frac{1}{2}|ab|$

Angular Bisectors of two lines:

The angular bisector of two intersecting lines is the locus of points which moves such that their distance from the lines is same.

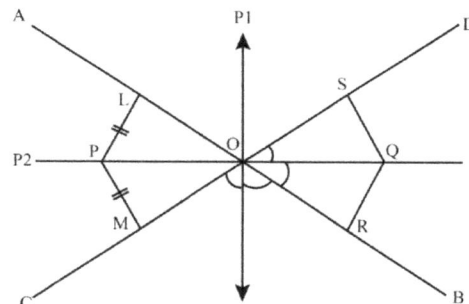

The equations of angular bisectors of the two intersecting lines $a_1 x + b_1 y + c_1 = 0$; $a_2 x + b_2 y + c_2 = 0$ are $\frac{a_1 x + b_1 y + c_1}{\sqrt{a_1^2 + b_1^2}} = \pm \frac{a_2 x + b_2 y + c_2}{\sqrt{a_2^2 + b_2^2}}$

Note:

(i) The equation of the acute angular bisector of two lines $a_1 x + b_1 y + c_1 = 0$; $a_2 x + b_2 y + c_2 = 0$ ($c_1 > 0$ and $c_2 > 0$) is $\frac{a_1 x + b_1 y + c_1}{\sqrt{a_1^2 + b_1^2}} = \frac{a_2 x + b_2 y + c_2}{\sqrt{a_2^2 + b_2^2}}$

when $a_1 a_2 + b_1 b_2 < 0$

(ii) The equation of the obtuse angular bisector of two lines $a_1x + b_1y + c_1 = 0$; $a_2x + b_2y + c_2 = 0$ ($c_1 > 0$ and $c_2 > 0$) is $\dfrac{a_1x+b_1y+c_1}{\sqrt{a_1^2+b_1^2}} = -\dfrac{a_2x+b_2y+c_2}{\sqrt{a_2^2+b_2^2}}$

when $a_1a_2 + b_1b_2 < 0$

(iii) The equation of the acute angular bisector of two lines $a_1x + b_1y + c_1 = 0$; $a_2x + b_2y + c_2 = 0$ ($c_1 > 0$ and $c_2 > 0$) is $\dfrac{a_1x+b_1y+c_1}{\sqrt{a_1^2+b_1^2}} = -\dfrac{a_2x+b_2y+c_2}{\sqrt{a_2^2+b_2^2}}$

when $a_1a_2 + b_1b_2 > 0$

(iv) The equation of the obtuse angular bisector of two lines $a_1x + b_1y + c_1 = 0$; $a_2x + b_2y + c_2 = 0$ ($c_1 > 0$ and $c_2 > 0$) is $\dfrac{a_1x+b_1y+c_1}{\sqrt{a_1^2+b_1^2}} = \dfrac{a_2x+b_2y+c_2}{\sqrt{a_2^2+b_2^2}}$

when $a_1a_2 + b_1b_2 > 0$

32. PAIR OF STRAIGHT LINES

Homogeneous equations:

Pair of lines: The combined equation of the pair of the straight lines which are passes through the origin is represented by the second-degree homogeneous equation

$ax^2 + 2hxy + by^2 = 0$

    (i) $ax^2 + 2hxy + by^2 = 0$ represents two lines $y = m_1 x$ and $y = m_2 x$ with

    $m_1 + m_2 = \frac{-2h}{b}$; $m_1 m_2 = \frac{a}{b}$

    (ii) $ax^2 + 2hxy + by^2 = 0$ represents two lines $l_1 x + m_1 y = 0$ and

    $l_2 x + m_2 y = 0$ with $l_1 l_2 = a$; $l_1 m_2 + l_2 m_1 = 2h$; $m_1 m_2 = b$

Nature of pair of lines: Let $S \equiv ax^2 + 2hxy + by^2 = 0$ represents

    (i) two real and distinct lines when $h^2 > ab$

    (ii) two coincident lines when $h^2 = ab$

    (iii) imaginary lines $h^2 < ab$

Angle between the pair of lines: The angle θ between the pair of lines represented by $ax^2 + 2hxy + by^2 = 0$ is given by $Cos\theta = \frac{|a+b|}{\sqrt{(a-b)^2+4h^2}}$

    (or) $Sin\theta = \frac{2\sqrt{h^2-ab}}{\sqrt{(a-b)^2+4h^2}}$

    (i) The two lines are perpendicular when $a + b = 0$

    (ii) The two lines are parallel or coincident when $h^2 = ab$

Pair of parallel and perpendicular lines:

    (i) The equation of pair of lines passes through the point (x_1, y_1) and parallel to the pair of lines $ax^2 + 2hxy + by^2 = 0$ is

    $a(x - x_1)^2 + 2h(x - x_1)(x - x_1) + b(y - y_1)^2 = 0$

    (ii) The equation of pair of lines passes through the point (x_1, y_1) and perpendicular to the pair of lines $ax^2 + 2hxy + by^2 = 0$ is

    $b(x - x_1)^2 - 2h(x - x_1)(x - x_1) + a(y - y_1)^2 = 0$

Product of the perpendiculars: The product of the perpendiculars from the point (α, β) to the pair of lines $ax^2 + 2hxy + by^2 = 0$ is $\frac{|a\alpha^2 + 2h\alpha\beta + b\beta^2|}{\sqrt{(a-b)^2 + 4h^2}}$

Centroid: If (α, β) is the centroid of the triangle whose sides $ax^2 + 2hxy + by^2 = 0$ is and $lx + my + n = 0$ then the centroid is given by

$$\frac{\alpha}{bl - hm} = \frac{\beta}{am - hl} = \frac{-2n}{3(bl^2 - 2hlm + am^2)}$$

Area of triangle:

(i) The area of the triangle formed by the straight line $lx + my + n = 0$ and the pair of lines $ax^2 + 2hxy + by^2 = 0$ is

$$\frac{n^2 \sqrt{h^2 - ab}}{|am^2 - 2hlm + bl^2|}$$

(ii) The area of the triangle formed by the straight line $lx + my + n = 0$ and the pair of lines $(lx + my)^2 - \tan^2\alpha(mx - ly)^2 = 0$ is

$$\frac{n^2}{\tan\alpha(l^2 + m^2)}$$

(iii) The area of the equilateral triangle formed by the straight line $lx + my + n = 0$ and the pair of lines $(lx + my)^2 - 3(mx - ly)^2 = 0$ is

$$\frac{n^2}{\sqrt{3}(l^2 + m^2)}$$

Pair of angular bisectors: The equation to the pair of bisectors of the angles between the pair of straight lines $ax^2 + 2hxy + by^2 = 0$ is

$$h(x^2 - y^2) = (a - b)xy$$

Non-Homogenous equations

Pair of lines: The combined equation of the pair of the straight lines which are not passes through the origin is represented by the second degree homogeneous equation

$$ax^2 + 2hxy + by^2 + 2gx + 2fy + c = 0$$

(i) $ax^2 + 2hxy + by^2 + 2gx + 2fy + c = 0$ represents two lines $y = m_1x + c_1$ and $y = m_2x + c_2$ with $m_1 + m_2 = \frac{-2h}{b}$; $m_1 m_2 = \frac{a}{b}$

(ii) $ax^2 + 2hxy + by^2 + 2gx + 2fy + c = 0$ represents two lines $l_1x + m_1y + n_1 = 0$ and $l_2x + m_2y + n_2 = 0$

with $l_1 l_2 = a$; $m_1 m_2 = b$; $n_1 n_2 = c$

$l_1 m_2 + l_2 m_1 = 2h$; $l_1 n_2 + l_2 n_1 = 2g$; $m_1 n_2 + m_2 n_1 = 2f$

Nature of pair of lines: Let $S \equiv ax^2 + 2hxy + by^2 + 2gx + 2fy + c = 0$ represents

(i) A pair of lines when $\Delta \equiv \begin{vmatrix} a & h & g \\ h & b & f \\ g & f & c \end{vmatrix} = abc + 2fgh - af^2 - bg^2 - ch^2 = 0$ and $h^2 \geq ab$; $g^2 \geq ac$; $f^2 \geq bc$

(ii) A pair of parallel lines when

$\Delta = 0$; $h^2 = ab$; $g^2 \geq ac$; $f^2 \geq bc$ and $af^2 = bg^2$

Distance between parallel lines: If $ax^2 + 2hxy + by^2 + 2gx + 2fy + c = 0$ represents a pair of parallel lines then the distance between the parallel lines is $2\sqrt{\frac{g^2-ac}{a(a+b)}}$ (or) $2\sqrt{\frac{f^2-bc}{b(a+b)}}$

Point of intersection of pair of lines: The point of intersection of the lines represented by

$S \equiv ax^2 + 2hxy + by^2 + 2gx + 2fy + c = 0$ is

$\left(\frac{hf-bg}{ab-h^2}, \frac{hg-af}{ab-h^2} \right)$ where $ab - h^2 \neq 0$

It will be obtain easily to solve the equations
$\frac{\partial S}{\partial x} = 0$ and $\frac{\partial S}{\partial y} = 0$

Angle between the pair of lines: The angle θ between the pair of lines represented by

$ax^2 + 2hxy + by^2 + 2gx + 2fy + c = 0$ is given by

$Cos\theta = \frac{|a+b|}{\sqrt{(a-b)^2+4h^2}}$ (or) $Sin\theta = \frac{2\sqrt{h^2-ab}}{\sqrt{(a-b)^2+4h^2}}$

Product of the perpendiculars:

(i) The product of the perpendiculars from the point (α, β) to the pair of lines

$ax^2 + 2hxy + by^2 + 2gx + 2fy + c = 0$ is

$\frac{|a\alpha^2 + 2h\alpha\beta + b\beta^2 + 2g\alpha + 2f\beta + c|}{\sqrt{(a-b)^2 + 4h^2}}$

(ii) The product of the perpendiculars from the point (0,0) to the pair of lines

$ax^2 + 2hxy + by^2 + 2gx + 2fy + c = 0$ is $\frac{|c|}{\sqrt{(a-b)^2+4h^2}}$

Area of triangle:

The area of the triangle formed by the straight line $lx + my + n = 0$ and the pair of lines $ax^2 + 2hxy + by^2 + 2gx + 2fy + c = 0$ whose point of intersection of the lines is (α, β) is

$$\frac{(l\alpha + m\beta + n)^2 \sqrt{h^2 - ab}}{|am^2 - 2hlm + bl^2|}$$

Pair of angular bisectors: If (α, β) is the point of intersection of the lines represented by $ax^2 + 2hxy + by^2 + 2gx + 2fy + c = 0$ then The equation to the pair of bisectors of the angles between the pair of straight lines is

$$h[(x - \alpha)^2 - (y - \beta)^2] = (a - b)(x - \alpha)(y - \beta)$$

Quadrilaterals formed by the both pair of lines:

The pair of lines represented by $ax^2 + 2hxy + by^2 = 0$ and the pair of lines represented by $ax^2 + 2hxy + by^2 + 2gx + 2fy + c = 0$ forms a

(i) Parallelogram when $a + b \neq 0$ and $(a - b)fg + h(f^2 - g^2) \neq 0$

(ii) Rhombus when $a + b \neq 0$ and $(a - b)fg + h(f^2 - g^2) = 0$

(iii) Rectangle when $a + b = 0$ and $(a - b)fg + h(f^2 - g^2) \neq 0$

(iv) Square when $a + b = 0$ and $(a - b)fg + h(f^2 - g^2) = 0$

Homogenisation: The process that joining of the points of intersection of a curve $S \equiv ax^2 + 2hxy + by^2 + 2gx + 2fy + c = 0$ and the line $L \equiv lx + my + n = 0$ to the origin is called as the homogenisation.

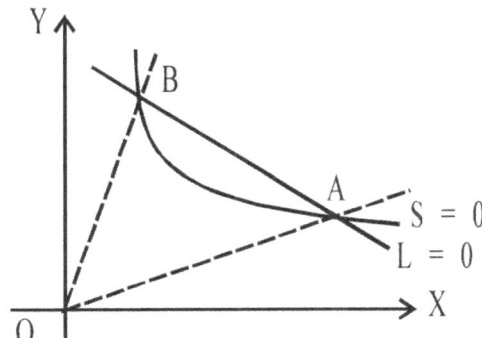

The equation of a curve $ax^2 + 2hxy + by^2 + 2gx + 2fy + c = 0$ is transformed as

$$ax^2 + 2hxy + by^2 + 2g\left(\frac{lx + my}{-n}\right) + 2f\left(\frac{lx + my}{-n}\right) + c\left(\frac{lx + my}{-n}\right)^2 = 0$$

33. CIRCLES

Circle: The circle is the locus of the points which are equidistance from a fixed point.

The fixed point is called as a center and it is denoted by C

The equidistance is called as a radius and it is denoted by r

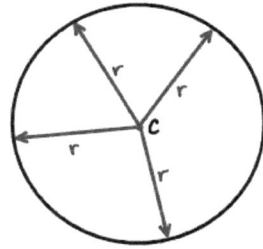

Equation of circle:

(i) The equation of circle having center (h, k) and radius r is

$(x - h)^2 + (y - k)^2 = r^2$

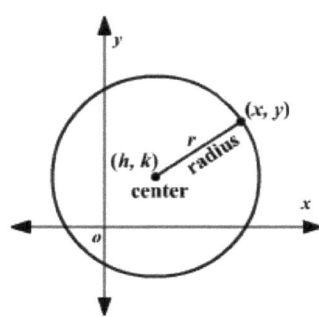

(ii) The equation of circle having center $(0,0)$ and radius r is $x^2 + y^2 = r^2$

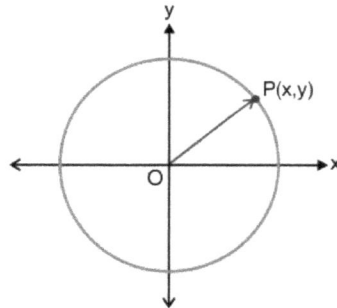

(iii) The general equation of circle is represented as $x^2 + y^2 + 2gx + 2fy + c = 0$ where center $C = (-g, -f)$ and radius $r = \sqrt{g^2 + f^2 - c}$

Second degree equation represents a circle:

The second-degree equation $ax^2 + 2hxy + by^2 + 2gx + 2fy + c = 0$ represents a circle when $a = b$ and $h = 0$

Then the second-degree equation represents a circle $ax^2 + ay^2 + 2gx + 2fy + c = 0$ where the center $C = \left(\frac{-g}{a}, \frac{-f}{a}\right)$ and radius $r = \frac{\sqrt{g^2+f^2-ac}}{|a|}$

Length of the intercepts:

(i) The length of the intercept made on X-axis by the circle

$x^2 + y^2 + 2gx + 2fy + c = 0$ is $2\sqrt{g^2 - c}$

(ii) The length of the intercept made on Y-axis by the circle

$x^2 + y^2 + 2gx + 2fy + c = 0$ is $2\sqrt{f^2 - c}$

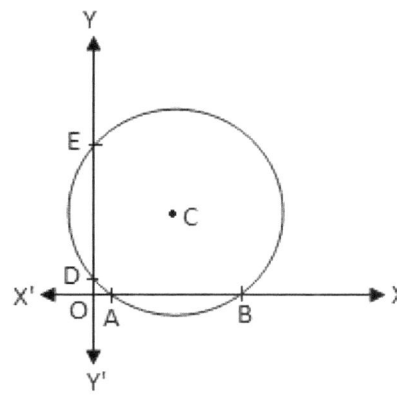

Here AB is the intercept made on X-axis and DE is the intercept made on Y-axis.

Conditions for circle touches the axes:

(i) If the circle touches the X-axis then $2\sqrt{g^2 - c} = 0$ i.e. $g = \pm c$

(ii) If the circle touches the Y-axis then $2\sqrt{f^2 - c} = 0$ i.e. $f = \pm c$

Equations of circles which touches the axes:

(i) The equation of circle which touches the X-axis is

$x^2 + y^2 \pm 2\sqrt{c}\, x + 2fy + c = 0$

(ii) The equation of circle which touches the Y-axis is

$x^2 + y^2 + 2gx \pm 2\sqrt{c}\, y + c = 0$

(iii) The equation of circle which touches the both the axes is

$x^2 + y^2 \pm 2\sqrt{c}\, x \pm 2\sqrt{c}\, y + c = 0$

Equation of circle whose extremities of diameter given:

The equation of a circle whose extremities of a diameter are $(x_1, y_1), (x_2, y_2)$ is
$(x - x_1)(x - x_2) + (y - y_1)(y - y_2) = 0$

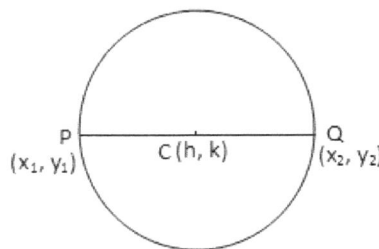

Parametric Equations:

(i) The parametric equations of the circle $x^2 + y^2 = r^2$ are
$$x = r \cos\theta \,; y = r \sin\theta$$

(ii) The parametric equations of the circle $(x - h)^2 + (y - k)^2 = r^2$ are
$$x = h \pm r \cos\theta \,; y = k \pm r \sin\theta$$

(iii) The parametric equations of the circle $x^2 + y^2 + 2gx + 2fy + c = 0$ are
$$x = -g \pm r \cos\theta \,; y = -f \pm r \sin\theta \text{ where } r = \sqrt{g^2 + f^2 - c}$$

Concentric circles: Two or more circles are said to be concentric circles when they have same center.

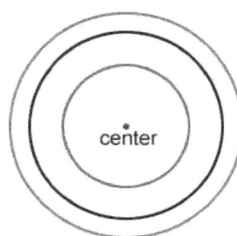

The equation of concentric circle of the circle $x^2 + y^2 + 2gx + 2fy + c = 0$ is of the form $x^2 + y^2 + 2gx + 2fy + k = 0$

Notations:

(i) $S \equiv x^2 + y^2 + 2gx + 2fy + c$

(ii) $S_1 = xx_1 + yy_1 + g(x+x_1) + f(y+y_1) + c$

(iii) $S_{11} \equiv x_1^2 + y_1^2 + 2gx_1 + 2fy_1 + c$

(iv) $S_{12} = x_1 x_2 + y_1 y_2 + g(x_1 + x_2) + f(y_1 + y_2) + c$

Power of a point: Let $S = 0$ be a circle and $P(x_1, y_1)$ be a point. Then the power of the point is represented by S_{11}

Position of a point w.r.to a circle: Let $S = 0$ be a circle and $P(x_1, y_1)$ be a point. Then

(i) P lies on the circle when $S_{11} = 0$

(ii) P lies inside of the circle when $S_{11} < 0$

(iii) P lies outside of the circle when $S_{11} > 0$

Length of the tangent: Let $S = 0$ be a circle and $P(x_1, y_1)$ be an outer point. If the tangent drawn from P meets the circle at A then PA is called the length of tangent and is given by

$PA = \sqrt{S_{11}}$

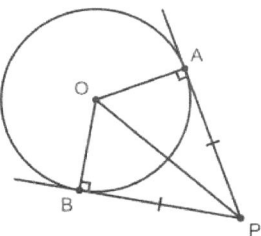

Equation of tangent: Let $S = 0$ be a circle and $P(x_1, y_1)$ be a point on the circle then the equation of the tangent is $S_1 = 0$

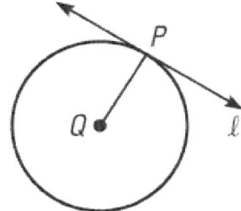

Equation of normal: Let $S = 0$ be a circle and $P(x_1, y_1)$ be a point on the circle then the equation of the normal is the line perpendicular to the tangent at P.

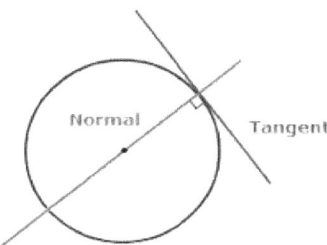

The equation of normal at $P(x_1, y_1)$ is $(x - x_1)(y_1 + f) - (y - y_1)(x_1 + g) = 0$

Normal is always passes through the center of the circle.

Equations of tangent and normal at parametric point:

(i) The equation of the tangent to the circle $x^2 + y^2 + 2gx + 2fy + c = 0$ at parametric point θ is $(x + g)\cos\theta + (y + f)\sin\theta = r$

(ii) The equation of the normal to the circle $x^2 + y^2 + 2gx + 2fy + c = 0$ at parametric point θ is $(x + g)\cos\theta = (y + f)\sin\theta$

Equation of Tangents having slope m: Let $S = 0$ be a circle with radius r and center $(-g, -f)$ then the equations of tangents with slope m are $y + f = m(x + g) \pm r\sqrt{1 + m^2}$

Condition for tangency: Let $S = 0$ be a circle with radius r and $L = 0$ be a straight line. Let d be the perpendicular distance drawn from the center to the line. Then

(i) The line is neither tough nor intersect the circle when $r < d$

(ii) The line is the tangent to the circle when $r = d$

(iii) The line is the secant to the circle when $r > d$

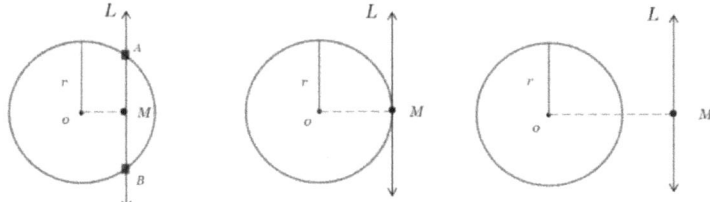

Point of contact: Let $S = 0$ be a circle with center C and $L = 0$ be a straight line. If the line touches the circle at P then the point of contact P is the foot of the perpendicular of C on the line.

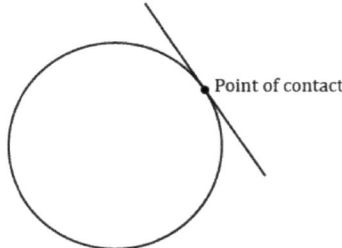

Length of the chord: Let $S = 0$ be a circle with radius r and $L = 0$ be a straight line. Let d be the perpendicular distance drawn from the center to the line. If the line intersects the circle at A and B then the length of the chord AB is given by $AB = 2\sqrt{r^2 - d^2}$

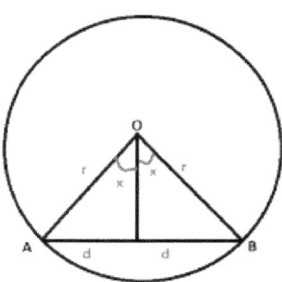

Equation of chord:

(i) The equation of the chord joining two points $(x_1, y_1), (x_2, y_2)$ on the circle $S = 0$ is $S_1 + S_2 = S_{12}$

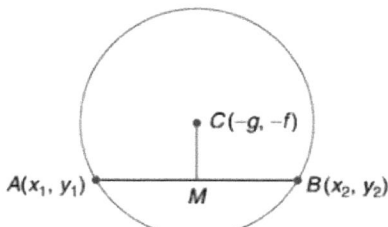

(ii) The equation of the chord whose midpoint is (x_1, y_1) to the circle $S = 0$ is $S_1 = S_{11}$

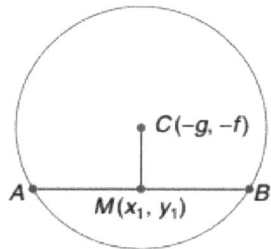

(iii) The equation of the chord joining two parametric points θ_1 and θ_2 on the circle $S = 0$ is $(x + g) \cos\left(\frac{\theta_1 + \theta_2}{2}\right) + (y + f) \sin\left(\frac{\theta_1 + \theta_2}{2}\right) = r \cos\left(\frac{\theta_1 - \theta_2}{2}\right)$ where r is the radius of the circle.

Chord of contact: Let $S = 0$ be a circle and $P(x_1, y_1)$ be an outer point. If the tangents drawn from P touches the circle at A and B then AB is called as the chord of contact of P.

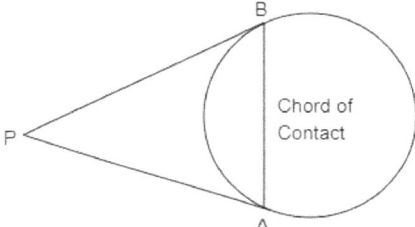

Equation of chord of contact: : Let $S = 0$ be a circle and $P(x_1, y_1)$ be an outer point. Then the equation of chord of contact of P is $S_1 = 0$

Pole and Polar: Let $S = 0$ be a circle and $P(x_1, y_1)$ be an outer point. Then the locus of points of intersections of tangents drawn at the intersection points of secants drawn from P is called as polar and P is called the pole.

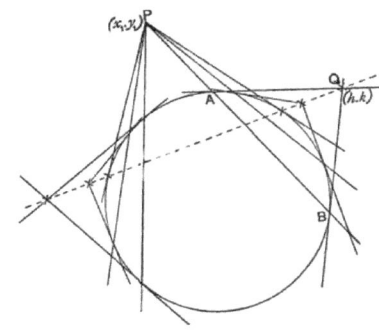

Equation of polar: Let $S = 0$ be a circle and $P(x_1, y_1)$ be an outer point. Then the equation of polar of P is $S_1 = 0$

Co-ordinates of pole: Let $x^2 + y^2 + 2gx + 2fy + c = 0$ be a circle and $lx + my + n = 0$ be the polar then the pole of the polar is

$$\left(-g + \frac{lr^2}{lg + mf - n}, -f + \frac{mr^2}{lg + mf - n}\right)$$

Conjugate points: Let $S = 0$ be a circle and $(x_1, y_1), (x_2, y_2)$ be two points. Then the points are said to be conjugate points if one point lie on the polar of another point.

The condition for the points $(x_1, y_1), (x_2, y_2)$ to be conjugate w.r.to the circle $S = 0$ is $S_{12} = 0$

Conjugate Lines: Let $S = 0$ be a circle and $l_1x + m_1y + n_1 = 0$; $l_2x + m_2y + n_2 = 0$ be two lines. Then the lines are said to be conjugate lines if pole of one line lie on the other line.

The condition for the lines to be conjugate w.r.to the circle is

$(l_1g + m_1f - n_1)(l_2g + m_2f - n_2) = r^2(l_1l_2 + m_1m_2)$

Equation to the pair of tangents: Let $S = 0$ be a circle and $P(x_1, y_1)$ be an outer point. Then the equation to the pair of tangents drawn from P to the circle is $S_1^2 = S.S_{11}$

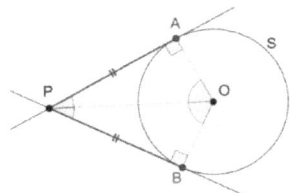

Angle between the pair of tangents: Let $S = 0$ be a circle and $P(x_1, y_1)$ be an outer point. If θ is the angle between the pair of tangents drawn from P then θ is given by $Tan\frac{\theta}{2} = \frac{r}{\sqrt{S_{11}}}$

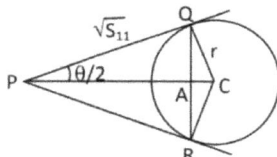

Area of the triangle made with pair of tangents:

Let $S = 0$ be a circle and $P(x_1, y_1)$ be an outer point. Then the area of the triangle made by the pair of tangents drawn from P and the chord of contact of P (in sq.units.) is $\frac{rS_{11}^{3/2}}{S_{11} + r^2}$

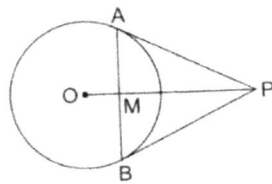

Length of chord of contact: Let $S = 0$ be a circle and $P(x_1, y_1)$ be an outer point. Let AB be the chord of contact of P. Then the length of AB is $2r\sqrt{\dfrac{S_{11}}{S_{11}+r^2}}$

Inverse point: Let $S = 0$ be a circle with center C radius r. Then the two points P and Q are said to be inverse points each other when

    (i) C, P, Q are collinear   **(ii)** P, Q lies same side of C   **(iii)** $CP.CQ = r^2$

Director Circle: The locus of the points of intersection of perpendicular tangents to a circle is called as a director circle.

The equation of the director circle to a circle $(x-h)^2 + (y-k)^2 = r^2$ is $(x-h)^2 + (y-k)^2 = 2r^2$

Longest and shortest distance of a point: Let $S = 0$ be a circle with center C and radius r and $P(x_1, y_1)$ be an outer point. Then

    (i) Longest distance from P to the circle = $CP + r$

    (ii) Shortest distance from P to the circle = $|CP - r|$

Relative positions of two circles:

Let C_1, C_2 be the centers and r_1, r_2 be the radii of two circles

SL. No	Condition	Diagram	Position	Number of Common tangents
1	$C_1C_2 > r_1 + r_2$		Two circles neither touch nor intersect	4
2	$C_1C_2 = r_1 + r_2$		Two circles touch each other externally	3
3	$|r_1 - r_2| < C_1C_2 < r_1 + r_2$		Two circles intersect each other	2
4	$C_1C_2 = |r_1 - r_2|$		Two circles touch each other internally	1
5	$C_1C_2 < |r_1 - r_2|$		One circle lies entirely in the other circle	0

Equation of common tangent: If two circles $S = 0$; $S^1 = 0$ touch each other then the equation of common tangent is $S - S^1 = 0$

Equation of common chord: If two circles $S = 0$; $S^1 = 0$ intersect each other then the equation of common chord is $S - S^1 = 0$

External Center of Similitude (E.C.S): The point of intersection of direct common tangents is called as external center of similitude.

Let $C_1(x_1, y_1), C_2(x_2, y_2)$ be the centers and r_1, r_2 be the radii of two circles.

Let $r_1 : r_2 = m : n$ Then E.C.S $= \left(\dfrac{mx_2 - nx_1}{m-n}, \dfrac{my_2 - ny_1}{m-n} \right)$

Internal Center of Similitude (I.C.S): The point of intersection of transverse common tangents is called as internal center of similitude.

Let $C_1(x_1, y_1), C_2(x_2, y_2)$ be the centers and r_1, r_2 be the radii of two circles.

Let $r_1 : r_2 = m : n$ Then I.C.S $= \left(\dfrac{mx_2 + nx_1}{m+n}, \dfrac{my_2 + ny_1}{m+n} \right)$

Length of common tangents:

Let d be the distance between the centres and r_1, r_2 be the radii of two circles. Then

    (i) The length of the direct common tangents $= \sqrt{d^2 - (r_1 - r_2)^2}$

    (ii) The length of the transverse common tangents $= \sqrt{d^2 - (r_1 + r_2)^2}$

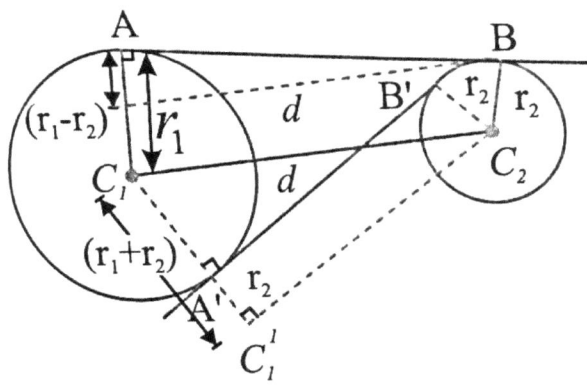

34. SYSTEM OF CIRCLES

Angle between two circles: The angle between two intersecting circles is the angle between the tangents drawn at intersecting points of the circles.

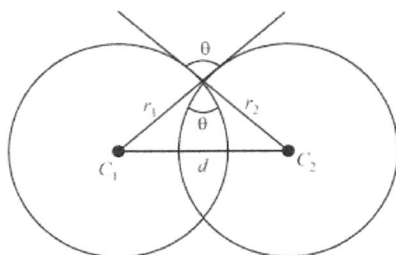

(i) Let $S = 0$; $S^1 = 0$ be two intersecting circles with centers C_1, C_2 and radii r_1, r_2 respectively and d is the distance between centers. Let θ be the angle between the two circles. Then θ is given by
$$Cos\ \theta = \frac{d^2 - r_1^2 - r_2^2}{2r_1 r_2}$$

(ii) Let $S \equiv x^2 + y^2 + 2gx + 2fy + c = 0$ and $S^1 \equiv x^2 + y^2 + 2g^1 x + 2f^1 y + c^1 = 0$ be two intersecting circles and θ be the angle between the two circles. Then θ is given by $Cos\ \theta = \dfrac{c + c^1 - 2gg^1 - 2ff^1}{2\sqrt{g^2 + f^2 - c}\sqrt{g^{1^2} + f^{1^2} - c^{1^2}}}$

Condition for orthogonality:

The two circles $S \equiv x^2 + y^2 + 2gx + 2fy + c = 0$ and $S^1 \equiv x^2 + y^2 + 2g^1 x + 2f^1 y + c^1 = 0$ are orthogonal then $2gg^1 + 2ff^1 = c + c^1$ or $d^2 = r_1^2 + r_2^2$ where d is the distance between the centers.

Common Chord: Let $S = 0$; $S^1 = 0$ be two intersecting circles then the equation of common chord is $S - S^1 = 0$

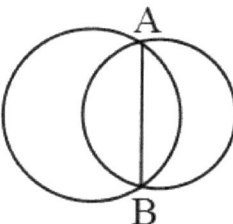

Length of common chord: If θ is the angle between the two circles whose radii r_1, r_2 then the length of the common chord is $\dfrac{2r_1 r_2\ Sin\theta}{\sqrt{r_1^2 + r_2^2 + 2r_1 r_2\ Cos\theta}}$

If the circles are orthogonal then the length of the chord is $\dfrac{2r_1 r_2}{\sqrt{r_1^2 + r_2^2}}$

Family of circles: Let $S = 0$; $S^1 = 0$ be two circles then the family of circles having $S = 0$; $S^1 = 0$ as members is $S + \lambda S^1 = 0$ or $S + \lambda L = 0$ where $L = S - S^1$

Radical axis: The locus of the points at which the length of the tangents drawn from these points to the two circles is equal is called as a radical axis of the circles.

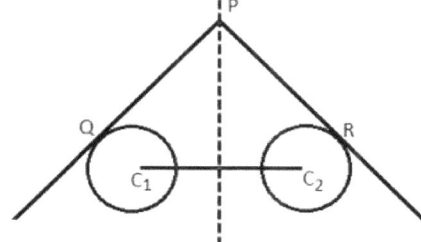

Equation of radical axis: Let $S = 0$; $S^1 = 0$ be two circles then the equation of radical axis of the two circles is $S - S^1 = 0$

Radical center: The radical center is the point of intersection of two radical axes formed by the three circles.

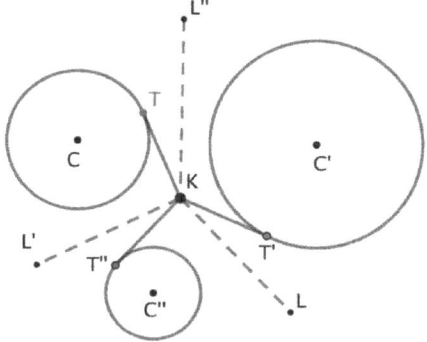

Let $S = 0$; $S^1 = 0$; $S^{11} = 0$ be three circles. Then the radical center is the point of intersecting of any two of $L \equiv S - S^1 = 0$; $L^1 \equiv S^1 - S^{11} = 0$; $L^{11} \equiv S - S^{11} = 0$

Properties of radical center:

(i) The length of tangent from radical center to any circle is same

(ii) If A, B, C are the centers of three circles which cut each other orthogonally then the radical center of the circles is the orthocenter of triangle ABC

(iii) If A, B, C are the centers of three circles which touch each other externally then the radical center of the circles is the incenter of triangle ABC

35. CONIC SECTION

Conic: The locus of the points which moves such that the ratio of their distances from a fixed point and fixed line is always a constant is called as a conic.

The fixed point is called as a focus and the fixed line is called the directrix.

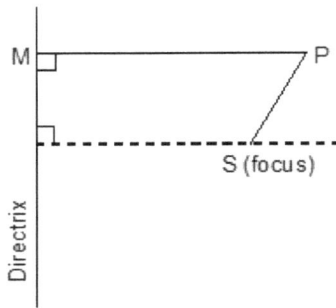

Let S be the focus and $L = 0$ be a fixed line. Let P be a point in the plane and PM be the perpendicular distance from P to the line. Then the locus of P is $\frac{SP}{PM} = e$ is a conic where e is a constant and it is called as eccentricity.

Note: (i) If $e = 1$ then the conic is a parabola

(ii) If $0 < e < 1$ then the conic is an ellipse

(iii) If $e > 1$ then the conic is a hyperbola

(iv) If $e = 0$ then the conic is a circle

(v) If $e \to \infty$ then the conic is a pair of straight lines

Conics represented by second degree equation:

Let $S \equiv ax^2 + 2hxy + by^2 + 2gx + 2fy + c = 0$ be a second degree equation and $\triangle \equiv abc + 2fgh - af^2 - bg^2 - ch^2$. Then

(i) $S = 0$ is a circle when $\triangle \neq 0, a = b \neq 0, h = 0$

(ii) $S = 0$ is a parabola when $\triangle \neq 0, h^2 = ab$

(iii) $S = 0$ is an ellipse (or) an empty set when $\triangle \neq 0, h^2 < ab$

(iv) $S = 0$ is a hyperbola when $\triangle \neq 0, h^2 > ab$

(v) $S = 0$ is a rectangular hyperbola when $\triangle \neq 0, h^2 > ab, a + b = 0$

Axis: The line passes through the focus and perpendicular to the directrix is called the axis of the conic

Vertex: The point of intersection of the conic and the axis is called as the vertex of the conic

Centre: The point which is equidistance from focus and the point of intersection of axis and directrix is called as a centre

Chord: The line segment joining any two points on the conic is called as the chord of the conic

Focal Chord: The chord passes through the focus is called as focal chord of the conic

Double ordinate: The chord which is perpendicular to the axis is called as a double ordinate

Latusrectum: The focal chord which is perpendicular to the axis is called as a latusrectum

Focal distance: The distance from any point on the conic from focus is called as a focal distance

36. PARABOLA

Parabola: The locus of the points which are equidistant from a fixed point and a fixed line is equal is called as a parabola.

Let S be the focus and $L = 0$ be a fixed line. Let P be a point in the plane and PM be the perpendicular distance from P to the line. Then the locus of P is $\frac{SP}{PM} = 1$ is a parabola

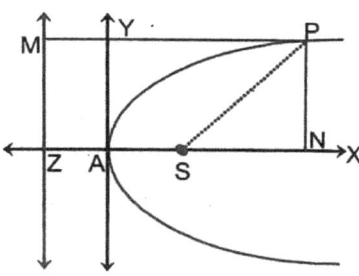

Various forms of parabola:

S.N.	Content	I	II	III	IV
1	Diagram				
2	Equation	$y^2 = 4ax$	$y^2 = -4ax$	$x^2 = 4ay$	$x^2 = -4ay$
3	Vertex	(0,0)	(0,0)	(0,0)	(0,0)
4	Focus	$(a, 0)$	$(-a, 0)$	$(0, a)$	$(0, -a)$
5	Eq. of Axis	$y = 0$	$y = 0$	$x = 0$	$x = 0$
6	Eq .of Directrix	$x = -a$	$x = a$	$y = -a$	$y = a$
7	Ends of Latusrectum	$(a, \pm 2a)$	$(-a, \pm 2a)$	$(\pm 2a, a)$	$(\pm 2a, -a)$
8	Length of Latusrectum	$4a$	$4a$	$4a$	$4a$

S.N	Content	V	VI	VII	VIII
1	Diagram	(y-k)² = 4a(x-h) with vertex A(h,k), focus S(h+a,k)	(y-k)² = -4a(x-h) with focus S(h-a,k), vertex A(h,k)	(x-h)² = 4a(y-k) with focus (h,k+a), vertex A(h,k)	(x-h)² = -4a(y-k) with vertex A(h,k), focus S(h,k-a)
2	Equation	$(y-k)^2 = 4a(x-h)$	$(y-k)^2 = -4a(x-h)$	$(x-h)^2 = 4a(y-k)$	$(x-h)^2 = -4a(y-k)$
3	Vertex	$(h+a, k)$	$(h-a, k)$	$(h, k+a)$	$(h, k-a)$
4	Focus	$(a, 0)$	$(-a, 0)$	$(0, a)$	$(0, -a)$
5	Eq. of Axis	$y = k$	$y = k$	$x = h$	$x = h$
6	Eq. of Directrix	$x = h - a$	$x = h + a$	$y = k - a$	$y = k + a$
7	Ends of Latusrectum	$(h+a, k \pm 2a)$	$(h-a, k \pm 2a)$	$(h \pm 2a, k+a)$	$(h \pm 2a, k-a)$
8	Length of Latusrectum	$4a$	$4a$	$4a$	$4a$

IX General equation of a parabola:

The equation of a parabola whose focus is (α, β) and the equation of directrix is $lx + my + n = 0$ is

$$(x - \alpha)^2 + (y - \beta)^2 = \frac{(l\alpha + m\beta + n)^2}{l^2 + m^2}$$

Equations of parabolas whose axes parallel to co-ordinate axes:

(i) The equation of a parabola whose axis is parallel to X-axis is

$x = ly^2 + my + n$ where $l, m, n \in \mathbb{R}$

Here, Length of the latusrectum is $\left| \frac{\text{Coeffecient of } x}{\text{Coefficeint of } y^2} \right|$

(ii) The equation of a parabola whose axis is parallel to X-axis is
$y = lx^2 + mx + n$ where $l, m, n \in \mathbb{R}$

Here, Length of the latusrectum is $\left|\dfrac{Coeffecient\ of\ y}{Coefficeint\ of\ x^2}\right|$

Notations:

(i) $S \equiv y^2 - 4ax$

(ii) $S_1 = yy_1 - 2a(x+x_1)$

(iii) $S_{11} \equiv y_1^2 - 4ax_1$

(iv) $S_{12} = y_1y_2 - 2a(x_1 + x_2)$

Parametric points:

(i) The parametric point of the parabola $y^2 = 4ax$ is $(at^2, 2at)$

(ii) The parametric point of the parabola $y^2 = -4ax$ is $(-at^2, 2at)$

(iii) The parametric point of the parabola $x^2 = 4ay$ is $(2at, at^2)$

(iv) The parametric point of the parabola $x^2 = -4ay$ is $(2at, -at^2)$

Position of a point w.r.to. parabola: Let $S = 0$ be a parabola and (x_1, y_1) be a point. Then

(i) The point lies on the parabola when $S_{11} = 0$

(ii) The point lies inside of the parabola when $S_{11} < 0$

(iii) The point lies outside of the parabola when $S_{11} > 0$

Equation of tangent:

(i) The equation of the tangent to $S = 0$ at the point (x_1, y_1) is $S_1 = 0$

(ii) The equation of tangent to $S = 0$ with slope m is $y = mx + \dfrac{a}{m}$

Equation of normal: The equation of normal is the line perpendicular to the tangent to the parabola.

(i) The equation of the normal to $S = 0$ at the point (x_1, y_1) is $y - y_1 = \dfrac{-y_1}{2a}(x - x_1)$

(ii) The equation of normal with slope m is $y = mx - 2am - am^3$

The equation has 3 values for m, so the number of normals can be drawn to the parabola is 3.

Equation of tangent and normal at the parametric point:

(i) The equation of tangent to $S = 0$ at the parametric point $t(at^2, 2at)$ is
$x - yt + at^2 = 0$

(ii) The equation of normal to $S = 0$ at the parametric point $t(at^2, 2at)$ is
$y + xt = 2at + at^3$

Condition for the tangency: Let $S = 0$ be a parabola and $y = mx + c$ be a line. Then

(i) The line is tangent to the parabola when $c = \frac{a}{m}$

(ii) The line intersects the parabola when $c > \frac{a}{m}$

(iii) The line neither touch nor intersect the parabola when $c < \frac{a}{m}$

The condition for the line $y = mx + c$ to be a tangent to the parabola $x^2 = 4ay$ is $c = -am^2$

Condition for a line normal to the parabola:

(i) If $y = mx + c$ is a normal to the parabola $S = 0$ then $c + 2am + am^3 = 0$

(ii) If $lx + my + n = 0$ is a normal to the parabola $S = 0$ then
$$al^3 + 2alm^2 + m^2n = 0$$

Number of normals: To find the number of normals, we have to find the value of $\triangle = G^2 + 4H^3$ where $G = \frac{-y_1}{a}$ and $H = \frac{2a-x_1}{3a}$

(i) If $\triangle > 0$ then the number of normals is 1

(ii) If $\triangle = 0$ then the number of normals is 2

(iii) If $\triangle < 0$ then the number of normals is 3

Point of contact:

(i) The point of contact of the parabola $y^2 = 4ax$ and the line $y = mx + c$ is $\left(\frac{a}{m^2}, \frac{2a}{m}\right)$

(ii) The point of contact of the parabola $y^2 = 4ax$ and the line $lx + my + n = 0$ is $\left(\frac{n}{l}, \frac{-2am}{l}\right)$

(iii) The point of contact of the parabola $x^2 = 4ay$ and the line $y = mx + c$ is $(2am, m^2)$

(iv) The point of contact of the parabola $x^2 = 4ay$ and the line $lx + my + n = 0$ is $\left(\frac{-2al}{m}, \frac{n}{m}\right)$

Focal distance: Let $y^2 = 4ax$ be a parabola and $P(x_1, y_1)$ be a point on the parabola then the focal distance is $SP = |x_1 + a|$

Focal chord: Let $y^2 = 4ax$ be a parabola and PSQ be a focal chord.

(i) If $P(at^2, 2at)$ be one end of the focal chord PSQ then the other end of the focal chord is $Q\left(\frac{a}{t^2}, \frac{-2a}{t}\right)$

(ii) If $P(at^2, 2at)$ be one end of the focal chord PSQ then the length of the focal chord is $a\left(t + \dfrac{1}{t}\right)^2$

(iii) If l is the semi laturectum then SP, l, SQ are in H.P. i.e.
$\dfrac{1}{SP} + \dfrac{1}{SQ} = \dfrac{2}{l}$

Equation of chord:

(i) The equation of the chord joining two points (x_1, y_1), (x_2, y_2) on the parabola $S = 0$ is $S_1 + S_2 = S_{12}$

(ii) The equation of the chord whose midpoint is (x_1, y_1) to the parabola $S = 0$ is $S_1 = S_{11}$

(iii) The equation of the chord joining two parametric points $(at_1^2, 2at_1)$ and $(at_2^2, 2at_2)$ on the parabola $S = 0$ is $y(t_1 + t_2) = 2x + 2at_1 t_2$

Length of chord:

(i) If the line $y = mx + c$ intersects the parabola $y^2 = 4ax$ in A and B then the length of the chord AB is $\dfrac{4}{m^2}\sqrt{a(a - mc)(1 + m^2)}$

(ii) The length of the chord joining the parametric points $(at_1^2, 2at_1)$ and $(at_2^2, 2at_2)$ on the parabola $S = 0$ is $a|t_1 - t_2|\sqrt{(t_1 + t_2)^2 + 4}$

(iii) The length of the chord having midpoint (x_1, y_1) to the parabola $S = 0$ is $\dfrac{1}{a}\sqrt{-S_{11}(y_1^2 + 4a^2)}$

Equation of chord of contact: The equation of chord of contact of $P(x_1, y_1)$ to the parabola $S = 0$ is $S_1 = 0$

The length of the chord of contact of (x_1, y_1) to the parabola $S = 0$ is $\dfrac{1}{a}\sqrt{S_{11}(y_1^2 + 4a^2)}$

Pair of tangents: The equation of pair of tangents drawn from $P(x_1, y_1)$ to the parabola $S = 0$ is $S_1^2 = S_{11}$

Angle between pair of tangents: If θ is the angle between the pair of tangents drawn from $P(x_1, y_1)$ to the parabola $S = 0$ then θ is given by
$Tan\theta = \dfrac{\sqrt{S_{11}}}{x_1 + a}$

Points of Intersection:

(i) The point of intersection of the tangents at t_1, t_2 on the parabola $S = 0$ is $(at_1 t_2, a(t_1 + t_2))$

(ii) The point of intersection of the normals at t_1, t_2 on the parabola $S = 0$ is $(2a + a\{(t_1 + t_2)^2 - t_1 t_2\}, at_1 t_2(t_1 + t_2))$

Area of triangles:

(i) The area of the triangle whose vertices (x_1, y_1), (x_2, y_2) and (x_3, y_3) inscribed in the parabola $S = 0$ is $\frac{1}{8a}|(y_1 - y_2)(y_2 - y_3)(y_3 - y_1)|$ sq. units

(ii) The area of the triangle formed by the tangents at three points (x_1, y_1), (x_2, y_2) and (x_3, y_3) on the parabola $S = 0$ is $\frac{1}{16a}|(y_1 - y_2)(y_2 - y_3)(y_3 - y_1)|$ sq. units

(iii) The area of the equilateral triangle inscribed in the parabola $S = 0$ with one vertex is the vertex of the parabola is $48\sqrt{3}a^2$ sq. units

Common Tangents:

(i) The equation of common tangent to the parabolas $y^2 = 4ax$ and $x^2 = 4ay$ is $xa^{1/3} + yb^{1/3} + a^{2/3}b^{2/3} = 0$

(ii) The equation of common tangent to the circle $x^2 + y^2 = 2a^2$ and the parabola $y^2 = 8ax$ is $y = \pm(x + 2a)$

Some more important points to remember:

(i) If the normal drawn at $(at_1^2, 2at_1)$ on the parabola $S = 0$ meets the parabola again at $(at_2^2, 2at_2)$ then $t_2 = -t_1 - \frac{2}{t_1}$

(ii) If two normal drawn at t_1, t_2 on the parabola $S = 0$ then $t_1 t_2 = -1$

(iii) The angle between the parabolas $y^2 = 4ax$ and $x^2 = 4ay$ is $Tan^{-1}\left[\frac{3a^{1/3}b^{1/3}}{2(a^{1/3} + b^{1/3})}\right]$

(iv) The tangents and normal drawn at the end points of latusrectum of the parabola $S = 0$ forms a square whose side is $2\sqrt{2}a$ and area $8a^2$ sq. units

(v) The orthocenter of the triangle formed be the points t_1, t_2, t_3 on the parabola $S = 0$ is $[-a, a(t_1 + t_2 + t_3 + t_1 t_2 t_3)]$

(vi) The angle between the tangents at t_1, t_2 on the parabola $S = 0$ is $Tan^{-1}\left|\frac{t_1 - t_2}{1 + t_1 t_2}\right|$

37. ELLIPSE

Ellipse: Let S be the focus and $L = 0$ be a fixed line. Let P be a point in the plane and PM be the perpendicular distance from P to the line. Then the locus of P is $\frac{SP}{PM} = e$ $(e < 1)$ is an ellipse where e is a constant and it is called as eccentricity.

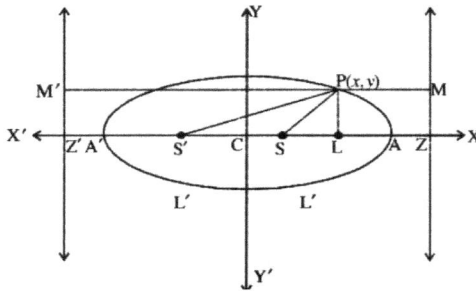

Various forms of ellipse:

S.N.	Content	I	II	III	IV
1	Diagram				
2	Equation	$\frac{x^2}{a^2} + \frac{y^2}{b^2} = 1$ $(a > b)$	$\frac{x^2}{a^2} + \frac{y^2}{b^2} = 1$ $(a < b)$	$\frac{(x-h)^2}{a^2} + \frac{(y-k)^2}{b^2} = 1$ $(a > b)$	$\frac{(x-h)^2}{a^2} + \frac{(y-k)^2}{b^2} = 1$ $(a < b)$
3	Center	$(0,0)$	$(0,0)$	(h,k)	(h,k)
4	Vertices	$(\pm a, 0)$	$(0, \pm b)$	$(h \pm a, k)$	$(h, k \pm b)$
5	Foci	$(\pm ae, 0)$	$(0, \pm be)$	$(h \pm ae, k)$	$(h, k \pm be)$
6	Length of Major Axis	$2a$	$2b$	$2a$	$2b$
7	Length of Minor Axis	$2b$	$2a$	$2b$	$2a$
8	Length of Latusrectum	$2b^2/a$	$2a^2/b$	$2b^2/a$	$2a^2/b$

9	Eccentricity	$\sqrt{\dfrac{a^2-b^2}{a^2}}$	$\sqrt{\dfrac{b^2-a^2}{b^2}}$	$\sqrt{\dfrac{a^2-b^2}{a^2}}$	$\sqrt{\dfrac{b^2-a^2}{b^2}}$
10	Distance between foci	$2ae$	$2be$	$2ae$	$2be$
11	Distance between directrices	$2a/e$	$2b/e$	$2a/e$	$2b/e$
12	Ends of Laturecta	$(\pm ae, \pm b^2/a)$	$(\pm a^2/b, \pm be)$	$(h \pm ae, k \pm b^2/a)$	$(h \pm a^2/b, k \pm be)$
13	Equations of directrices	$x = \pm a/e$	$y = \pm b/e$	$x = h \pm a/e$	$y = k \pm b/e$

General equation of an Ellipse:

The equation of an ellipse whose focus is (α, β) and the equation of directrix is $lx + my + n = 0$ is

$$(x - \alpha)^2 + (y - \beta)^2 = e^2 \frac{(l\alpha + m\beta + n)^2}{l^2 + m^2} \text{ where } e < 1$$

Notations:

(i) $S \equiv \dfrac{x^2}{a^2} + \dfrac{y^2}{b^2} - 1 = 0 \; (a > b)$

(ii) $S_1 = \dfrac{xx_1}{a^2} + \dfrac{yy_1}{b^2} - 1 = 0$

(iii) $S_{11} \equiv \dfrac{x_1^2}{a^2} + \dfrac{y_1^2}{b^2} - 1 = 0$

(iv) $S_{12} = \dfrac{x_1 x_2}{a^2} + \dfrac{y_1 y_2}{b^2} - 1 = 0$

Parametric points:

(i) The parametric point of the ellipse $\dfrac{x^2}{a^2} + \dfrac{y^2}{b^2} = 1$ is $(a\cos\theta, b\sin\theta)$

(ii) The parametric point of the ellipse $\dfrac{(x-h)^2}{a^2} + \dfrac{(y-k)^2}{b^2} = 1$ is $(h + a\cos\theta, k + b\sin\theta)$

Position of a point w.r.to. Ellipse: Let $S = 0$ be an ellipse and (x_1, y_1) be a point. Then

(i) The point lies on the ellipse when $S_{11} = 0$

(ii) The point lies inside of the ellipse when $S_{11} < 0$

(iii) The point lies outside of the ellipse when $S_{11} > 0$

Equation of tangent:

(i) The equation of the tangent to $S = 0$ at the point (x_1, y_1) is $S_1 = 0$

i.e. $\frac{xx_1}{a^2} + \frac{yy_1}{b^2} = 1$

(ii) The equation of tangent to $S = 0$ with slope m is $y = mx \pm \sqrt{a^2m^2 + b^2}$

Equation of normal:
The equation of normal is the line perpendicular to the tangent to the parabola.

(i) The equation of the normal to $S = 0$ at the point (x_1, y_1) is

$$\frac{a^2 x}{x_1} - \frac{b^2 y}{y_1} = a^2 - b^2$$

(ii) The equation of normal with slope m is

$$y = mx \pm \frac{m(a^2 - b^2)}{\sqrt{a^2 + m^2 b^2}}$$

Equation of tangent and normal at the parametric point:

(i) The equation of the tangent at θ to the ellipse $S = 0$ is

$$\frac{x \cos\theta}{a} + \frac{y \sin\theta}{b} = 1$$

(ii) The equation of the normal at θ to the ellipse $S = 0$ is

$$\frac{ax}{\cos\theta} - \frac{by}{\sin\theta} = a^2 - b^2$$

Condition for the tangency:

(i) If the line $y = mx + c$ is a tangent to the ellipse $\frac{x^2}{a^2} + \frac{y^2}{b^2} = 1$ is $c^2 = a^2m^2 + b^2$

(ii) If the line $lx + my + n = 0$ is a tangent to the ellipse $\frac{x^2}{a^2} + \frac{y^2}{b^2} = 1$ is $a^2 l^2 + b^2 m^2 = n^2$

Condition for a line normal to the Ellipse:

(i) If the line $y = mx + c$ is a normal to the ellipse $\frac{x^2}{a^2} + \frac{y^2}{b^2} = 1$ is

$$c^2 = \frac{m^2(a^2 - b^2)^2}{a^2 + b^2 m^2}$$

(ii) If the line $lx + my + n = 0$ is a tangent to the ellipse $\frac{x^2}{a^2} + \frac{y^2}{b^2} = 1$ is

$$\frac{a^2}{l^2} + \frac{b^2}{m^2} = \frac{(a^2 - b^2)^2}{n^2}$$

Point of contact:

(i) If the line $y = mx + c$ is a tangent to the ellipse $\frac{x^2}{a^2} + \frac{y^2}{b^2} = 1$ then the point of contact is $\left(\frac{-a^2m}{c}, \frac{-b^2m}{c}\right)$

(ii) If the line $lx + my + n = 0$ is a tangent to the ellipse $\frac{x^2}{a^2} + \frac{y^2}{b^2} = 1$ then the point of contact is $\left(\frac{-a^2l}{n}, \frac{-b^2m}{n}\right)$

Focal distance: Let $\frac{x^2}{a^2} + \frac{y^2}{b^2} = 1$ be an ellipse and $P(x_1, y_1)$ be a point on the ellipse then the focal distance is $SP = |x_1 + a|$

The sum of the focal distances is equal to the length of the major axis

i.e. $SP + S^1P = 2a$

Focal chord: Let $\frac{x^2}{a^2} + \frac{y^2}{b^2} = 1$ be an ellipse and PSQ be a focal chord. Then for semi latusrectum l

$$\frac{1}{SP} + \frac{1}{SQ} = \frac{2}{l}$$

Equation of chord:

(i) The equation of the chord joining two points $(x_1, y_1), (x_2, y_2)$ on the ellipse

$S = 0$ is $S_1 + S_2 = S_{12}$

(ii) The equation of the chord whose midpoint is (x_1, y_1) to the ellipse

$S = 0$ is $S_1 = S_{11}$

(iii) The equation of the chord joining two parametric points α and β is

$$\frac{x}{a} \cos\left(\frac{\alpha + \beta}{2}\right) + \frac{y}{b} \sin\left(\frac{\alpha + \beta}{2}\right) = \cos\left(\frac{\alpha - \beta}{2}\right)$$

Equation of chord of contact: The equation of chord of contact of $P(x_1, y_1)$ to the ellipse $S = 0$ is $S_1 = 0$

Pair of tangents: The equation of pair of tangents drawn from $P(x_1, y_1)$ to the ellipse $S = 0$ is $S_1^2 = S_{11}$

Angle between pair of tangents: If θ is the angle between the pair of tangents drawn from $P(x_1, y_1)$ to the ellipse $S = 0$ then θ is given by $Tan\theta = \left|\frac{2ab\sqrt{S_{11}}}{x_1^2 + y_1^2 - a^2 - b^2}\right|$

Director circle: The locus of the points of intersection of perpendicular tangents is called as a director circle

(i) The equation of the director circle to the ellipse $\frac{x^2}{a^2} + \frac{y^2}{b^2} = 1$ is $x^2 + y^2 = a^2 + b^2$

(ii) The equation of the director circle to the ellipse $\frac{(x-h)^2}{a^2} + \frac{(y-k)^2}{b^2} = 1$ is $(x-h)^2 + (y-k)^2 = a^2 + b^2$

Auxiliary circle; The locus of the feet of the perpendiculars drawn from foci to any tangent to the ellipse is called as an auxiliary circle.

The equation of the director circle to the ellipse $\frac{x^2}{a^2} + \frac{y^2}{b^2} = 1$ is $x^2 + y^2 = a^2$ if $a > b$ or $x^2 + y^2 = b^2$ if $a < b$

Equation of the ellipse referred to two perpendicular lines:

If in a ellipse, the equation and its length of major axis are $lx + my + n_1 = 0$, $2a$ and the equation and its length of minor axis are $mx - ly + n_2 = 0$, $2b$ then the equation of the ellipse is $\dfrac{\left(\frac{mx-ly+n_2}{\sqrt{l^2+m^2}}\right)^2}{a^2} + \dfrac{\left(\frac{lx+my+n_1}{\sqrt{l^2+m^2}}\right)^2}{b^2} = 1$

38. HYPERBOLA

Hyperbola: Let S be the focus and $L = 0$ be a fixed line. Let P be a point in the plane and PM be the perpendicular distance from P to the line. Then the locus of P is $\frac{SP}{PM} = e$ $(e > 1)$ is an hyperbola where e is a constant and it is called as eccentricity.

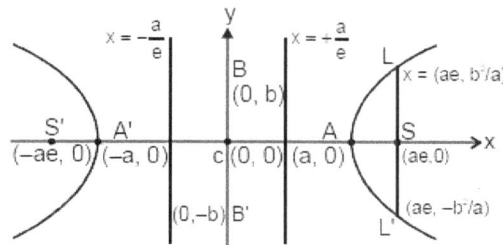

Various forms of Hyperbola:

S.N.	Content	I	II	III	IV
1	Diagram				
2	Equation	$\dfrac{x^2}{a^2} - \dfrac{y^2}{b^2} = 1$	$\dfrac{x^2}{a^2} - \dfrac{y^2}{b^2} = -1$	$\dfrac{(x-h)^2}{a^2} - \dfrac{(y-k)^2}{b^2} = 1$	$\dfrac{(x-h)^2}{a^2} - \dfrac{(y-k)^2}{b^2} = -1$
3	Center	$(0,0)$	$(0,0)$	(h,k)	(h,k)
4	Vertices	$(\pm a, 0)$	$(0, \pm b)$	$(h \pm a, k)$	$(h, k \pm b)$
5	Foci	$(\pm ae, 0)$	$(0, \pm be)$	$(h \pm ae, k)$	$(h, k \pm be)$
6	Length of transverse Axis	$2a$	$2b$	$2a$	$2b$
7	Length of conjugate Axis	$2b$	$2a$	$2b$	$2a$
8	Length of Latusrectum	$2b^2/a$	$2a^2/b$	$2b^2/a$	$2a^2/b$

9	Eccentricity	$\sqrt{\dfrac{a^2+b^2}{a^2}}$	$\sqrt{\dfrac{a^2+b^2}{b^2}}$	$\sqrt{\dfrac{a^2+b^2}{a^2}}$	$\sqrt{\dfrac{a^2+b^2}{b^2}}$
10	Distance between foci	$2ae$	$2be$	$2ae$	$2be$
11	Distance between directrices	$2a/e$	$2b/e$	$2a/e$	$2b/e$
12	Ends of Laturecta	$(\pm ae, \pm b^2/a)$	$(\pm a^2/b, \pm be)$	$(h \pm ae, k \pm b^2/a)$	$(h \pm a^2/b, k \pm be)$
13	Equations of directrices	$x = \pm a/e$	$y = \pm b/e$	$x = h \pm a/e$	$y = k \pm b/e$

General equation of a hyperbola:

The equation of a hyperbola whose focus is (α, β) and the equation of directrix is $lx + my + n = 0$ is

$(x - \alpha)^2 + (y - \beta)^2 = e^2 \dfrac{(l\alpha + m\beta + n)^2}{l^2 + m^2}$ where $e > 1$

Conjugate Hyperbola: The equation of conjugate hyperbola to the hyperbola $\dfrac{x^2}{a^2} - \dfrac{y^2}{b^2} = 1$ is $\dfrac{x^2}{a^2} - \dfrac{y^2}{b^2} = -1$

If e_1 and e_2 are the eccentricities of hyperbola and conjugate hyperbola then

$\dfrac{1}{e_1^2} + \dfrac{1}{e_2^2} = 1$

Notations:

(i) $S \equiv \dfrac{x^2}{a^2} - \dfrac{y^2}{b^2} - 1 = 0$

(ii) $S_1 = \dfrac{xx_1}{a^2} - \dfrac{yy_1}{b^2} - 1 = 0$

(iii) $S_{11} \equiv \dfrac{x_1^2}{a^2} - \dfrac{y_1^2}{b^2} - 1 = 0$

(iv) $S_{12} = \dfrac{x_1 x_2}{a^2} - \dfrac{y_1 y_2}{b^2} - 1 = 0$

Parametric points:

(i) The parametric point of the hyperbola $\frac{x^2}{a^2} - \frac{y^2}{b^2} = 1$ is $(a\,Sec\theta, b\,Tan\theta)$

(ii) The parametric point of the hyperbola $\frac{(x-h)^2}{a^2} - \frac{(y-k)^2}{b^2} = 1$ is $(h + a\,Sec\theta, k + b\,Tan\theta)$

Position of a point w.r.to. Hyperbola:
Let $S = 0$ be a hyperbola and (x_1, y_1) be a point. Then

(i) The point lies on the hyperbola when $S_{11} = 0$

(ii) The point lies inside of the hyperbola when $S_{11} < 0$

(iii) The point lies outside of the hyperbola when $S_{11} > 0$

Equation of tangent:

(i) The equation of the tangent to $S = 0$ at the point (x_1, y_1) is $S_1 = 0$ i.e.
$$\frac{xx_1}{a^2} - \frac{yy_1}{b^2} = 1$$

(ii) The equation of tangent to $S = 0$ with slope m is $y = mx \pm \sqrt{a^2m^2 - b^2}$

Equation of normal:
The equation of normal is the line perpendicular to the tangent to the parabola.

(i) The equation of the normal to $S = 0$ at the point (x_1, y_1) is
$$\frac{a^2x}{x_1} + \frac{b^2y}{y_1} = a^2 + b^2$$

(ii) The equation of normal with slope m is
$$y = mx \pm \frac{m(a^2 + b^2)}{a^2 - b^2}$$

Equation of tangent and normal at the parametric point:

(i) The equation of the tangent at θ to the hyperbola $S = 0$ is
$$\frac{x\,Sec\theta}{a} - \frac{y\,Tan\theta}{b} = 1$$

(ii) The equation of the normal at θ to the hyperbola $S = 0$ is
$$\frac{ax}{Sec\theta} + \frac{by}{Tan\theta} = a^2 + b^2$$

Condition for the tangency:

(i) If the line $y = mx + c$ is a tangent to the hyperbola $\frac{x^2}{a^2} - \frac{y^2}{b^2} = 1$ is
$$c^2 = a^2m^2 - b^2$$

(ii) If the line $lx + my + n = 0$ is a tangent to the hyperbola $\frac{x^2}{a^2} - \frac{y^2}{b^2} = 1$ is $a^2l^2 - b^2m^2 = n^2$

Condition for a line normal to the Hyperbola:

(i) If the line $y = mx + c$ is a normal to the hyperbola $\frac{x^2}{a^2} - \frac{y^2}{b^2} = 1$ is
$$c^2 = \frac{m^2(a^2+b^2)^2}{(a^2-b^2)^2}$$

(ii) If the line $lx + my + n = 0$ is a tangent to the hyperbola $\frac{x^2}{a^2} - \frac{y^2}{b^2} = 1$ is
$$\frac{a^2}{l^2} - \frac{b^2}{m^2} = \frac{(a^2+b^2)^2}{n^2}$$

Point of contact:

(i) If the line $y = mx + c$ is a tangent to the hyperbola $\frac{x^2}{a^2} - \frac{y^2}{b^2} = 1$ then the point of contact is $\left(\frac{-a^2m}{c}, \frac{-b^2}{c}\right)$

(ii) If the line $lx + my + n = 0$ is a tangent to the hyperbola $\frac{x^2}{a^2} - \frac{y^2}{b^2} = 1$ then the point of contact is $\left(\frac{-a^2l}{n}, \frac{b^2m}{n}\right)$

Focal distance: Let $\frac{x^2}{a^2} - \frac{y^2}{b^2} = 1$ be a hyperbola and $P(x_1, y_1)$ be a point on the hyperbola then the focal distance is $SP = |ex_1 - a|$

Focal chord: Let $\frac{x^2}{a^2} - \frac{y^2}{b^2} = 1$ be a hyperbola and PSQ be a focal chord. Then for semi latusrectum l
$$\frac{1}{SP} + \frac{1}{SQ} = \frac{2}{l}$$

Equation of chord:

(i) The equation of the chord joining two points $(x_1, y_1), (x_2, y_2)$ on the hyperbola $S = 0$ is $S_1 + S_2 = S_{12}$

(ii) The equation of the chord whose midpoint is (x_1, y_1) to the hyperbola $S = 0$ is $S_1 = S_{11}$

(iii) The equation of the chord joining two parametric points α and β is
$$\frac{x}{a} Cos\left(\frac{\alpha - \beta}{2}\right) - \frac{y}{b} Sin\left(\frac{\alpha + \beta}{2}\right) = Cos\left(\frac{\alpha + \beta}{2}\right)$$

Equation of chord of contact: The equation of chord of contact of $P(x_1, y_1)$ to the hyperbola $S = 0$ is $S_1 = 0$

Pair of tangents: The equation of pair of tangents drawn from $P(x_1, y_1)$ to the hyperbola $S = 0$ is $S_1^2 = S_{11}$

Angle between pair of tangents: If θ is the angle between the pair of tangents drawn from $P(x_1, y_1)$ to the hyperbola $S = 0$ then θ is given by

$$Tan\theta = \left| \frac{2ab\sqrt{-S_{11}}}{x_1^2 + y_1^2 - a^2 + b^2} \right|$$

Director circle: The locus of the points of intersection of perpendicular tangents is called as a director circle

(i) The equation of the director circle to the hyperbola $\frac{x^2}{a^2} - \frac{y^2}{b^2} = 1$ is

$x^2 + y^2 = a^2 - b^2$

(ii) The equation of the director circle to the hyperbola $\frac{(x-h)^2}{a^2} - \frac{(y-k)^2}{b^2} = 1$

Is $(x - h)^2 + (y - k)^2 = a^2 - b^2$

Auxiliary circle: The locus of the feet of the perpendiculars drawn from foci to any tangent to the ellipse is called as an auxiliary circle.

The equation of the director circle to the hyperbola $\frac{x^2}{a^2} - \frac{y^2}{b^2} = 1$ is $x^2 + y^2 = a^2$

Asymptotes: The tangents at infinity to the hyperbola are called as asymptotes.

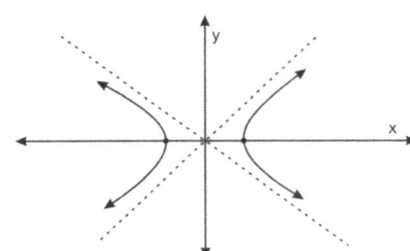

(i) The pair of equations of asymptotes of hyperbola $S = 0$ is $\frac{x^2}{a^2} - \frac{y^2}{b^2} = 0$

(ii) The asymptotes always passes through the center

(iii) The product of the perpendiculars from any point on the hyperbola to the asymptotes is $\frac{a^2 b^2}{a^2 + b^2}$

(iv) The angle between the asymptotes is $2\,Sec^{-1}(e)$ or $2\,Tan^{-1}\left(\frac{b}{a}\right)$

(v) The asymptotes of hyperbola and the asymptotes of conjugate hyperbola are same

Rectangular Hyperbola $x^2 - y^2 = a^2$: The hyperbola in which the length of the transverse and conjugate axes are equal is called as a rectangular hyperbola.

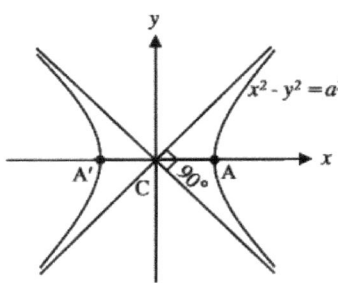

(i) Center=(0,0) ; Vertices=$(\pm a, 0)$ Foci=$(\pm ae, 0)$; eccentricity (e)=$\sqrt{2}$

(ii) Parametric point is $(a\ Sec\theta, aTan\theta)$

(iii) The equations of asymptotes are $y = \pm x$

(iv) The angle between the asymptotes is 90^0

Rectangular Hyperbola $xy = c^2$: If we rotate the rectangular hyperbola $x^2 - y^2 = a^2$ through an angle of 45^0 in anticlockwise direction then it become $xy = c^2$

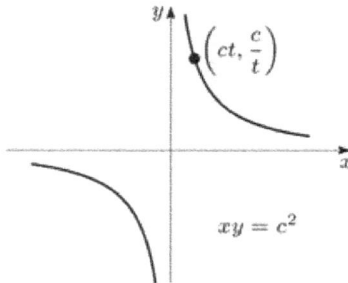

(i) Center=(0,0) ; Vertices=$(\pm c, \pm c)$; Foci=$(\pm\sqrt{2}c, \pm\sqrt{2}c)$;

eccentricity (e)=$\sqrt{2}$

(ii) Parametric point is $\left(ct, \dfrac{c}{t}\right)$

(iii) The equations of asymptotes are co-ordinate axes

(iv) The angle between the asymptotes is 90^0

Equation of the hyperbola referred to two perpendicular lines:

If in a hyperbola, the equation and its length of transverse axis are $lx + my + n_1 = 0$, $2a$ and the equation and its length of conjugate axis are $mx - ly + n_2 = 0$, $2b$ then the equation of the hyperbola is $\dfrac{\left(\dfrac{mx-ly+n_2}{\sqrt{l^2+m^2}}\right)^2}{a^2} - \dfrac{\left(\dfrac{lx+my+n_1}{\sqrt{l^2+m^2}}\right)^2}{b^2} = 1$

3D GEOMETRY

39. 3D CO-ORDINATES

Rectangular cartesian coordinate system: Let $XOX^1; YOY^1$ and ZOZ^1 be three mutually perpendicular lines intersecting at origin O. The system formed with these lines is called as rectangular cartesian coordinate system.

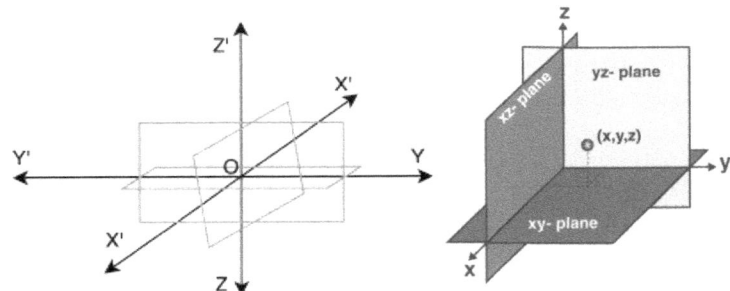

(i) $XOX^1; YOY^1$ and ZOZ^1 are called as X-axis, Y-axis and ZZ-axis respectively

(ii) The planes denoted by XOX^1, YOY^1; YOY^1, ZOZ^1; and XOX^1, ZOZ^1 are called as XY-plane, YZ-plane and ZX-plane respectively

3D coordinates: Let P be any point in the space. Let PM be the perpendicular from P to the XY-plane. Let MN be the perpendicular from M to the X-axis such that MN parallel to Y-axis. Then ON, MN, PM are called as x, y, z co-ordinates of P

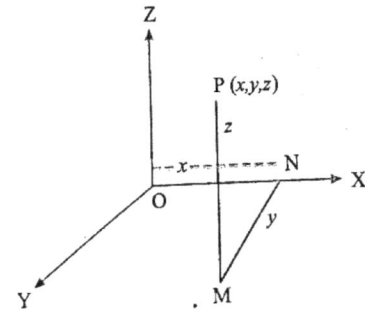

If ON=x, MN=y, PM=z then the coordinates of P are (x,y,z)

3D Octants:

Octant	$OXYZ$	OX^1YZ	OX^1Y^1Z	OXY^1Z	$OXYZ^1$	OX^1YZ^1	$OX^1Y^1Z^1$	OXY^1Z^1
	I	II	III	IV	V	VI	VII	VIII
x	+	−	−	+	+	−	−	+
y	+	+	−	−	+	+	−	−
z	+	+	+	+	−	−	−	−

Distance between two points:

(i) The distance between the two points $A(x_1, y_1, z_1)$ and $B(x_2, y_2, z_2)$ is

$$AB = \sqrt{(x_2 - x_1)^2 + (y_2 - y_1)^2 + (z_2 - z_1)^2}$$

(ii) The distance between the two points $O(0,0)$ and $A(x, y, z)$ is

$$OA = \sqrt{x^2 + y^2 + z^2}$$

Perpendicular distances:

The perpendicular distance of the point $P(x, y, z)$ from

(i) X-axis is $\sqrt{y^2 + z^2}$ (ii) Y-axis is $\sqrt{x^2 + z^2}$ (iii) Z-axis is $\sqrt{x^2 + y^2}$

(iv) XY-plane is $|z|$ (v) YZ-plane is $|x|$ (vi) ZX-plane is $|y|$

Collinear Points: Three are more points are said to be collinear when they lie on a same line.

To show $A(x_1, y_1, z_1), B(x_2, y_2, z_2)$ and $C(x_3, y_3, z_3)$ are collinear, we have to show that

$$\frac{x_1 - x_2}{x_2 - x_3} = \frac{y_1 - y_2}{y_2 - y_3} = \frac{z_1 - z_2}{z_2 - z_3} \text{ or } \begin{vmatrix} x_1 & y_1 & z_1 \\ x_2 & y_2 & z_2 \\ x_3 & y_3 & z_3 \end{vmatrix} = 0$$

Section Formula:

(i) If a point P divides the line joining the points $A(x_1, y_1, z_1), B(x_2, y_2, z_2)$ in the ratio $m:n$ internally then

$$P = \left(\frac{mx_2 + nx_1}{m + n}, \frac{my_2 + ny_1}{m + n}, \frac{mz_2 + nz_1}{m + n} \right)$$

(ii) If a point Q divides the line joining the points $A(x_1, y_1, z_1), B(x_2, y_2, z_2)$ in the ratio $m:n$ externally then

$$Q = \left(\frac{mx_2 - nx_1}{m - n}, \frac{my_2 - ny_1}{m - n}, \frac{mz_2 - nz_1}{m - n} \right)$$

Planes divides the line segment: Let the line segment joining the points $A(x_1, y_1, z_1), B(x_2, y_2, z_2)$ then

(i) XY-plane divides AB in the ratio $-z_1 : z_2$

(ii) YZ-plane divides AB in the ratio $-x_1 : x_2$

(iii) ZX-plane divides AB in the ratio $-y_1 : y_2$

Midpoint: The midpoint of the line segment joining the points $A(x_1, y_1, z_1), B(x_2, y_2, z_2)$ is

$$\left(\frac{x_1 + x_2}{2}, \frac{y_1 + y_2}{2}, \frac{z_1 + z_2}{2} \right)$$

Points of Trisection: The points which divides the line segment joining two points in the ratio $1:2$ or $2:1$ are called as the points of trisection.

Harmonic Conjugate: If the points P and Q divides the line segment joining tow points in the same ratio internally and externally respectively then one of P and Q is called as harmonic conjugate to the other.

Centroid of the triangle:

The centroid of the triangle formed by the points $A(x_1, y_1, z_1), B(x_2, y_2, z_2)$ and $C(x_3, y_3, z_3)$ is $G = \left(\frac{x_1+x_2+x_3}{3}, \frac{y_1+y_2+y_3}{3}, \frac{z_1+z_2+z_3}{3}\right)$

Centroid of the tetrahedron:

The centroid of the tetrahedron formed by the points $A(x_1, y_1, z_1), B(x_2, y_2, z_2)$, $C(x_3, y_3, z_3)$ and $D(x_4, y_4, z_4)$ is $G = \left(\frac{x_1+x_2+x_3+x_4}{4}, \frac{y_1+y_2+y_3+y_4}{4}, \frac{z_1+z_2+z_3+z_4}{4}\right)$

Relation between centroid, orthocentre and circumcentre:

Let centroid, orthocentre and circumcentre of a triangle be denoted by G, O, S respectively. Then G, O, S are collinear and G divides the line segment joining O and S in the ratio 2:1 internally. i.e. OG:GS=2:1

Translation of axes: If the origin is shifted to (h, k, l) without rotate the angle of axes then the point (x, y, z) changes to (X, Y, Z)

$x = X + h \,; y = Y + k \,; z = Z + l$

40. DIRECTION COSINES AND DIRECTION RATIOS

Direction cosines of a line: If a line L passing through the origin makes angles α, β, γ with the positive X, Y, Z-axes respectively then $Cos\alpha, Cos\beta, Cos\gamma$ are called as the direction cosines of the line.

Generally, the direction cosines represented by l, m, n.

Relations:

(i) $l^2 + m^2 + n^2 = 1$

(ii) $Cos^2\alpha + Cos^2\beta + Cos^2\gamma = 1$

(iii) $Sin^2\alpha + Sin^2\beta + Sin^2\gamma = 2$

Direction ratios of a line: If a line passes through the points (x_1, y_1, z_1), $B(x_2, y_2, z_2)$ then the direction ratios of the line represented as $\pm k(x_2 - x_1, y_2 - y_1, z_2 - z_1)$ for $k \in \mathbb{R}$

Then the direction cosines represented as $\pm k \left(\frac{x_2 - x_1}{AB}, \frac{y_2 - y_1}{AB}, \frac{z_2 - z_1}{AB} \right)$ for $k \in \mathbb{R}$

If (a, b, c) are the direction ratios and (l, m, n) are the direction cosines then

$$\frac{l}{a} = \frac{m}{b} = \frac{n}{c} = \pm \frac{1}{\sqrt{a^2 + b^2 + c^2}}$$

Angle between two lines:

(i) Let (l_1, m_1, n_1), (l_2, m_2, n_2) be the direction cosines of two lines. If θ is the angle between two lines then

$$Cos\theta = |l_1 l_2 + m_1 m_2 + n_1 n_2| \quad (or) \quad Sin\theta = \sqrt{\sum (l_1 m_2 - l_2 m_1)^2}$$

(ii) Let (a_1, b_1, c_1), (a_2, b_2, c_2) be the direction ratios of two lines. If θ is the angle between two lines then

$$Cos\theta = \frac{|a_1 a_2 + b_1 b_2 + c_1 c_2|}{\sqrt{a_1^2 + b_1^2 + c_1^2} \sqrt{a_2^2 + b_2^2 + c_2^2}}$$

Conditions for the lines to be parallel or perpendicular:

(a) Let (l_1, m_1, n_1), (l_2, m_2, n_2) be the direction cosines of two lines. Then

(i) The lines are perpendicular when $l_1 l_2 + m_1 m_2 + n_1 n_2 = 0$

(ii) The lines are parallel when $\frac{l_1}{l_2} = \frac{m_1}{m_2} = \frac{n_1}{n_2}$

(b) Let (a_1, b_1, c_1), (a_2, b_2, c_2) be the direction ratios of two lines. Then

  (i) The lines are perpendicular when $a_1a_2 + b_1b_2 + c_1c_2 = 0$

  (ii) The lines are parallel when $\frac{a_1}{a_2} = \frac{b_1}{b_2} = \frac{c_1}{c_2}$

Connected by the relations:

(a) If the direction cosines (l, m, n) of two lines are connected by the relations $al + bm + cn = 0$ and $umn + vnl + wlm = 0$ then

  (i) The lines are perpendicular when $\frac{u}{a} + \frac{v}{b} + \frac{w}{c} = 0$

  (ii) The lines are parallel when $\sqrt{au} \pm \sqrt{bv} \pm \sqrt{cw} = 0$

(b) If the direction cosines (l, m, n) of two lines are connected by the relations $al + bm + cn = 0$ and $ul^2 + vm^2 + wn^2 = 0$ then

  (i) The lines are perpendicular when $\sum a^2(v + w) = 0$

  (ii) The lines are parallel when $\frac{a^2}{u} + \frac{b^2}{v} + \frac{c^2}{w} = 0$

Important points related to cube:

  (i) The angle between any two diagonals of a cube is $Cos^{-1}\left(\frac{1}{3}\right)$

  (ii) The angle between a diagonal and the diagonal of a face of a cube is

$$Cos^{-1}\sqrt{\frac{2}{3}}$$

  (iii) If a line makes aangles $\alpha, \beta, \gamma, \delta$ with the four diagonals of a cube then

$$Cos^2\alpha + Cos^2\beta + Cos^2\gamma + Cos^2\delta = \frac{4}{3} \; ; \; Sin^2\alpha + Sin^2\beta + Sin^2\gamma + Sin^2\delta = \frac{8}{3}$$

Length of the projection:

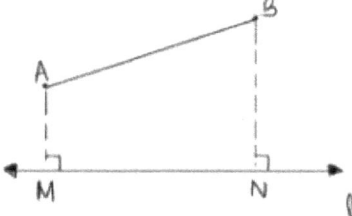

Let A, B be two points. Let l be the line. Let M, N be the projections of A and B respectively on the line. Let θ be the angle made by AB with the line. Then MN is the projection of AB on the line and is given by $MN = AB\, Cos\theta$

41. 3D PLANES

General equation of plane: The general equation of the plane is represented as $ax + by + cz + d = 0$

Equations of planes:

(i) The equation of the plane passing through the point (x_1, y_1, z_1) and having d.r's of normal as (a, b, c) is $a(x - x_1) + b(y - y_1) + c(z - z_1) = 0$

(ii) The equation of the plane passing through the point (x_1, y_1, z_1) and parallel to the plane $ax + by + cz + d = 0$ is

$a(x - x_1) + b(y - y_1) + c(z - z_1) = 0$

(iii) The equation of the plane passing through the point (x_1, y_1, z_1) and parallel to the lines having d.r's are (a_1, b_1, c_1) and (a_2, b_2, c_2) is

$$\begin{vmatrix} x - x_1 & y - y_1 & z - z_1 \\ a_1 & b_1 & c_1 \\ a_2 & b_2 & c_2 \end{vmatrix} = 0$$

(iv) The equation of the plane passing through the points (x_1, y_1, z_1), (x_2, y_2, z_2) and parallel to the line having d.r.'s are (a, b, c) is

$$\begin{vmatrix} x - x_1 & y - y_1 & z - z_1 \\ x_2 - x_1 & y_2 - y_1 & z_2 - z_1 \\ a & b & c \end{vmatrix} = 0$$

(v) The equation of the plane passing through the points (x_1, y_1, z_1), (x_2, y_2, z_2) and (x_3, y_3, z_3) is $\begin{vmatrix} x - x_1 & y - y_1 & z - z_1 \\ x_2 - x_1 & y_2 - y_1 & z_2 - z_1 \\ x_3 - x_1 & y_3 - y_1 & z_3 - z_1 \end{vmatrix} = 0$

Intercept form: If a plane cuts X, Y, Z-axes at the points A, B, C respectively then the equation of the plane in intercept form is $\frac{x}{a} + \frac{y}{b} + \frac{z}{c} = 1$ where X-intercept is a, Y-intercept is b and Z-intercept is c.

Here $A = (a, 0, 0); B = (0, b, 0); C = (0, 0, c)$

Normal form:

(i) If (l, m, n) bare the direction cosines of a normal to the plane and p is the perpendicular distance from origin to the plane then the equation of the plane is

$lx + my + nz = p$ where $l^2 + m^2 + n^2 = 1$

(ii) The normal form of the plane $ax + by + cz + d = 0$ is

$$\frac{a}{\sqrt{a^2 + b^2 + c^2}} + \frac{b}{\sqrt{a^2 + b^2 + c^2}} + \frac{c}{\sqrt{a^2 + b^2 + c^2}} = \frac{-d}{\sqrt{a^2 + b^2 + c^2}} \quad (d < 0)$$

$$\frac{-a}{\sqrt{a^2 + b^2 + c^2}} + \frac{-b}{\sqrt{a^2 + b^2 + c^2}} + \frac{-c}{\sqrt{a^2 + b^2 + c^2}} = \frac{d}{\sqrt{a^2 + b^2 + c^2}} \quad (d > 0)$$

Perpendicular distances:

(i) The perpendicular distance from a point (x_1, y_1, z_1) to the plane

$ax + by + cz + d = 0$ is $\frac{|ax_1+by_1+cz_1+d|}{\sqrt{a^2+b^2+c^2}}$

(ii) The perpendicular distance from a point $(0,0,0)$ to the plane

$ax + by + cz + d = 0$ is $\frac{|d|}{\sqrt{a^2+b^2+c^2}}$

(iii) The distance between the planes

$a_1x + b_1y + c_1z + d_1 = 0$ and $a_2x + b_2y + c_2z + d_2 = 0$ is $\frac{|d_1-d_2|}{\sqrt{a^2+b^2+c^2}}$

Angle between two planes: If θ is the angle between two planes $a_1x + b_1y + c_1z + d_1 = 0$ and $a_2x + b_2y + c_2z + d_2 = 0$ then $Cos\theta = \frac{|a_1a_2+b_1b_2+c_1c_2|}{\sqrt{a_1^2+b_1^2+c_1^2}\sqrt{a_2^2+b_2^2+c_2^2}}$

Conditions for the planes to be parallel or perpendicular:

Let $a_1x + b_1y + c_1z + d_1 = 0$ and $a_2x + b_2y + c_2z + d_2 = 0$ be the two planes. Then

(i) The planes are perpendicular when $a_1a_2 + b_1b_2 + c_1c_2 = 0$

(ii) The planes are parallel when $\frac{a_1}{a_2} = \frac{b_1}{b_2} = \frac{c_1}{c_2}$

Foot of the perpendicular:

If $Q(h, k, l)$ is the foot of the perpendicular drawn from the point $P(x_1, y_1, z_1)$ to the plane $ax + by + cz + d = 0$ then

$$\frac{h-x_1}{a} = \frac{k-y_1}{b} = \frac{l-z_1}{c} = \frac{-(ax_1+by_1+cz_1+d)}{a^2+b^2+c^2}$$

Image: If $Q(h, k, l)$ is the image of the point $P(x_1, y_1, z_1)$ w.r.to the plane

$ax + by + cz + d = 0$ then

$$\frac{h-x_1}{a} = \frac{k-y_1}{b} = \frac{l-z_1}{c} = \frac{-2(ax_1+by_1+cz_1+d)}{a^2+b^2+c^2}$$

Ratio formed by the plane: The ratio in which the plane $\pi \equiv ax + by + cz + d = 0$ divides the line segment joining the points $(x_1, y_1, z_1), (x_2, y_2, z_2)$ is $-\pi_1 : \pi_2$

Position of a point w.r.to to a plane: Let $\pi \equiv ax + by + cz + d = 0$ be a plane and $(x_1, y_1, z_1), (x_2, y_2, z_2)$ be two points. Then

(i) The points lie on same side of the plane when $\pi_1 : \pi_2 > 0$

(ii) The points lie on opposite side of the plane when $\pi_1 : \pi_2 < 0$

Equations to the pair of angular bisectors of the planes:

The equations of angular bisectors of the two intersecting planes
$a_1x + b_1y + c_1z + d_1 = 0$; $a_2x + b_2y + c_2z + d_2 = 0$ are

$$\frac{a_1x + b_1y + c_1z + d_1}{\sqrt{a_1^2 + b_1^2 + c_1^2}} = \pm \frac{a_2x + b_2y + c_2z + d_2}{\sqrt{a_2^2 + b_2^2 + c_2^2}}$$

Note:

(i) The equation of the acute angular bisector of two planes
$a_1x + b_1y + c_1z + d_1 = 0$; $a_2x + b_2y + c_2z + d_2 = 0$ $(d_1 > 0 \text{ and } d_2 > 0)$ is
$$\frac{a_1x + b_1y + c_1z + d_1}{\sqrt{a_1^2 + b_1^2 + c_1^2}} = \frac{a_2x + b_2y + c_2z + d_2}{\sqrt{a_2^2 + b_2^2 + c_2^2}} \quad \text{when } a_1a_2 + b_1b_2 + c_1c_2 < 0$$

(ii) The equation of the obtuse angular bisector of two planes
$a_1x + b_1y + c_1z + d_1 = 0$; $a_2x + b_2y + c_2z + d_2 = 0$ $(d_1 > 0 \text{ and } d_2 > 0)$ is
$$\frac{a_1x + b_1y + c_1z + d_1}{\sqrt{a_1^2 + b_1^2 + c_1^2}} = -\frac{a_2x + b_2y + c_2z + d_2}{\sqrt{a_2^2 + b_2^2 + c_2^2}} \quad \text{when } a_1a_2 + b_1b_2 + c_1c_2 < 0$$

(iii) The equation of the acute angular bisector of two planes
$a_1x + b_1y + c_1z + d_1 = 0$; $a_2x + b_2y + c_2z + d_2 = 0$ $(d_1 > 0 \text{ and } d_2 > 0)$ is
$$\frac{a_1x + b_1y + c_1z + d_1}{\sqrt{a_1^2 + b_1^2 + c_1^2}} = -\frac{a_2x + b_2y + c_2z + d_2}{\sqrt{a_2^2 + b_2^2 + c_2^2}} \quad \text{when } a_1a_2 + b_1b_2 + c_1c_2 > 0$$

(iv) The equation of the obtuse angular bisector of two planes
$a_1x + b_1y + c_1z + d_1 = 0$; $a_2x + b_2y + c_2z + d_2 = 0$ $(d_1 > 0 \text{ and } d_2 > 0)$ is
$$\frac{a_1x + b_1y + c_1z + d_1}{\sqrt{a_1^2 + b_1^2 + c_1^2}} = \frac{a_2x + b_2y + c_2z + d_2}{\sqrt{a_2^2 + b_2^2 + c_2^2}} \quad \text{when } a_1a_2 + b_1b_2 + c_1c_2 > 0$$

42. 3D LINES

3D line: The intersection of two planes is a line.

Equation of a line:

(i) The equation of a line passing through (x_1, y_1, z_1) and having d. c's (l, m, n) is $\dfrac{x-x_1}{l} = \dfrac{y-y_1}{m} = \dfrac{z-z_1}{n}$

(ii) The equation of a line passing through (x_1, y_1, z_1) and having d.r's (a, b, c) is $\dfrac{x-x_1}{a} = \dfrac{y-y_1}{b} = \dfrac{z-z_1}{c}$

(iii) The equation of a line passing through the points (x_1, y_1, z_1) and (x_2, y_2, z_2) is $\dfrac{x-x_1}{x_2-x_1} = \dfrac{y-y_1}{y_2-y_1} = \dfrac{z-z_1}{z_2-z_1}$

Parametric form of the line: If the line passes through the point $P(x_1, y_1, z_1)$ and having d.c's (l, m, n) then the parametric form of the line is $x = x_1 + lr$; $y = y_1 + mr$; $z = z_1 + nr$ where $r = OP$

Angle between to lines: If θ is the angle between the lines $\dfrac{x-x_1}{a_1} = \dfrac{y-y_1}{b_1} = \dfrac{z-z_1}{c_1}$ and

$\dfrac{x-x_2}{a_2} = \dfrac{y-y_2}{b_2} = \dfrac{z-z_2}{c_2}$ then $Cos\theta = \dfrac{|a_1 a_2 + b_1 b_2 + c_1 c_2|}{\sqrt{a_1^2+b_1^2+c_1^2}\sqrt{a_2^2+b_2^2+c_2^2}}$

Conditions for the lines to be parallel or perpendicular:

(i) The planes are perpendicular when $a_1 a_2 + b_1 b_2 + c_1 c_2 = 0$

(ii) The planes are parallel when $\dfrac{a_1}{a_2} = \dfrac{b_1}{b_2} = \dfrac{c_1}{c_2}$

Angle between a line and a plane: If θ is the angle between the line $\dfrac{x-x_1}{l} = \dfrac{y-y_1}{m} = \dfrac{z-z_1}{n}$ and the plane $ax + by + cz + d = 0$ then

$Sin\theta = \dfrac{|al + bm + cn|}{\sqrt{a^2 + b^2 + c^2}\sqrt{l^2 + m^2 + n^2}}$

(i) The planes are perpendicular when $\dfrac{l}{a} = \dfrac{m}{b} = \dfrac{n}{c}$

(ii) The planes are parallel when $al + bm + cn = 0$

Coplanar lines: Two lines are said to be coplanar when they are either parallel or intersect.

Equation of a plane containing the lines: The equation of the plane containing the lines

$\frac{x-x_1}{a_1} = \frac{y-y_1}{b_1} = \frac{z-z_1}{c_1}$ and $\frac{x-x_2}{a_2} = \frac{y-y_2}{b_2} = \frac{z-z_2}{c_2}$ is

$$\begin{vmatrix} x-x_1 & y-y_1 & z-z_1 \\ a_1 & b_1 & c_1 \\ a_2 & b_2 & c_2 \end{vmatrix} = 0 \text{ or } \begin{vmatrix} x-x_2 & y-y_2 & z-z_2 \\ a_1 & b_1 & c_1 \\ a_2 & b_2 & c_2 \end{vmatrix} = 0$$

Condition for two lines are coplanar:

(i) The line $\frac{x-x_1}{l} = \frac{y-y_1}{m} = \frac{z-z_1}{n}$ lies in the plane $ax + by + cz + d = 0$ when $ax_1 + by_1 + cz_1 + d = 0$ and $al + bm + cn = 0$

(ii) The lines $\frac{x-x_1}{a_1} = \frac{y-y_1}{b_1} = \frac{z-z_1}{c_1}$ and $\frac{x-x_2}{a_2} = \frac{y-y_2}{b_2} = \frac{z-z_2}{c_2}$ are coplanar when

$$\begin{vmatrix} x_1-x_2 & y_1-y_2 & z_1-z_2 \\ a_1 & b_1 & c_1 \\ a_2 & b_2 & c_2 \end{vmatrix} = 0$$

(iii) The lines $\frac{x-x_1}{l} = \frac{y-y_1}{m} = \frac{z-z_1}{n}$ and $a_1x + b_1y + c_1z + d_1 = a_2x + b_2y + c_2z + d_2 = 0$ are coplanar then

$$\frac{a_1x_1 + b_1y_1 + c_1z_1 + d_1}{a_1l + b_1m + c_1n} = \frac{a_2x_1 + b_2y_1 + c_2z_1 + d_2}{a_2l + b_2m + c_2n}$$

Skew lines (Non-Coplanar lines): Two lines are said to be skew lines when they are neither parallel nor intersecting.

Shortest distance between skew lines:

The shortest distance between skew lines

$\frac{x-x_1}{a_1} = \frac{y-y_1}{b_1} = \frac{z-z_1}{c_1}$ and $\frac{x-x_2}{a_2} = \frac{y-y_2}{b_2} = \frac{z-z_2}{c_2}$ is

$$\frac{\begin{vmatrix} x_1-x_2 & y_1-y_2 & z_1-z_2 \\ a_1 & b_1 & c_1 \\ a_2 & b_2 & c_2 \end{vmatrix}}{\sqrt{\sum(a_1b_2 - a_2b_1)^2}}$$

Distance between parallel lines: The distance between the parallel lines

$\frac{x-x_1}{a} = \frac{y-y_1}{b} = \frac{z-z_1}{c}$ and $\frac{x-x_2}{a} = \frac{y-y_2}{b} = \frac{z-z_2}{c}$ is $\left| \frac{\begin{vmatrix} \vec{i} & \vec{j} & \vec{k} \\ a & b & c \\ x_1-x_2 & y_1-y_2 & z_1-z_2 \end{vmatrix}}{\sqrt{a^2+b^2+c^2}} \right|$

VECTOR ALGEBRA

43. ADDITION OF VECTORS

Scalar: A physical quantity which has only magnitude is called as a scalar

    Eg: Work, length, speed etc.,

Vector: A physical quantity which has both magnitude and direction is called as a vector

    Eg: Torque, force, velocity etc.,

Notation of a vector: Let A be the initial point and B be the terminal point. Then the vector is denoted by $\overline{AB}$. If $AB = a$ then the vector is also denoted by $\bar{a}$.

Types of vectors

(i) Position vector: Let O be a fixed point and P be any point in the space. Then $\overline{OP}$ is called as a position vector of P.

(ii) Co-initial vectors: Two are more vectors are said to be co-initial vectors when they have same initial point

(iii) Co-terminal vectors: Two are more vectors are said to be co-terminal vectors when they have same terminal point

(iv) Null vector: If the magnitude of a vector is zero then the vector is called as a null vector

(v) Unit vector: If the magnitude of a vector is one then the vector is called as an unit vector

    (a) The unit vector in the direction of $\bar{a}$ is $\dfrac{\bar{a}}{|\bar{a}|}$

    (b) The unit vector in the opposite direction of $\bar{a}$ is $-\dfrac{\bar{a}}{|\bar{a}|}$

    (c) The unit vector parallel to the vector $\bar{a}$ is $\pm\dfrac{\bar{a}}{|\bar{a}|}$

    (d) The unit vector parallel to the vector $\bar{a}$ and having magnitude k units is $\pm k\dfrac{\bar{a}}{|\bar{a}|}$

Equal vectors: Two vectors $\bar{a}$ and $\bar{b}$ are called equal when they have same magnitude and direction.

Addition of vectors:

(i) Triangle law of vectors: Let $\overline{AB} = \bar{a}$ and $\overline{BC} = \bar{b}$ be two non-zero vectors represented by two sides of a triangle ABC then the resultant vector is given by the side $\overline{AC}$

i.e. $\overline{AC} = \overline{AB} + \overline{BC} = \bar{a} + \bar{b}$

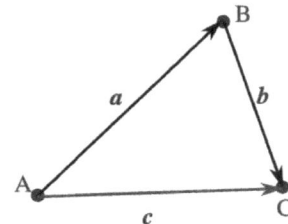

(ii) Parallelogram law of vectors: Let $\bar{a}$ and $\bar{b}$ be the adjacent sides of a parallelogram then their resultant $\bar{a} + \bar{b}$ represents the diagonal of the parallelogram through the common points.

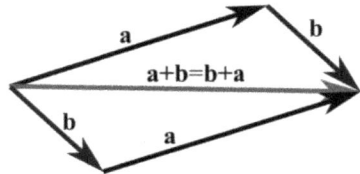

Properties of addition of vectors:

(i) **Commutative law:** $\bar{a} + \bar{b} = \bar{b} + \bar{a}$

(ii) **Associative law:** $\bar{a} + (\bar{b} + \bar{c}) = (\bar{a} + \bar{b}) + \bar{c}$

(iii) **Existence of identity:** $\bar{a} + \bar{0} = \bar{0} + \bar{a} = \bar{a}$

(iv) **Existence of inverse:** $\bar{a} + (-\bar{a}) = (-\bar{a}) + \bar{a} = \bar{0}$

Note: (i) $|\bar{a} + \bar{b}| \leq |\bar{a}| + |\bar{b}|$ (ii) $|\bar{a} - \bar{b}| \leq |\bar{a}| + |\bar{b}|$

(iii) $|\bar{a} - \bar{b}| \geq ||\bar{a}| - |\bar{b}||$

Scalar multiplication of vectors: Let $\bar{a}$ be a vector and k be a scalar then the scalar multiplication of the vector is denoted by $k\bar{a}$ that means the vector multiplied by the scalar k.

Properties of Scalar multiplication of vectors: Let $\bar{a}$, $\bar{b}$ be the two vectors and l, m be the scalars. Then

(i) $(l + m)\bar{a} = l\bar{a} + m\bar{a}$

(ii) $l(\bar{a} + \bar{b}) = l\bar{a} + l\bar{b}$

(iii) $l(m\bar{a}) = m(l\bar{a}) = lm\bar{a}$

Representation of a vector in 3D: Let $\bar{i}, \bar{j}, \bar{k}$ be the unit vectors acting along the positive directions of X, Y, Z-axes respectively. Then the position of vector of P is denoted as $\overline{OP} = x\bar{i} + y\bar{j} + z\bar{k}$ and the point P is in 3-dimensional plane denoted as (x, y, z).

Here $|\overline{OP}| = \sqrt{x^2 + y^2 + z^2}$

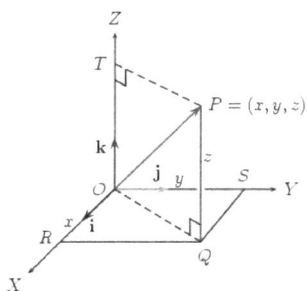

Right handed and left handed systems:

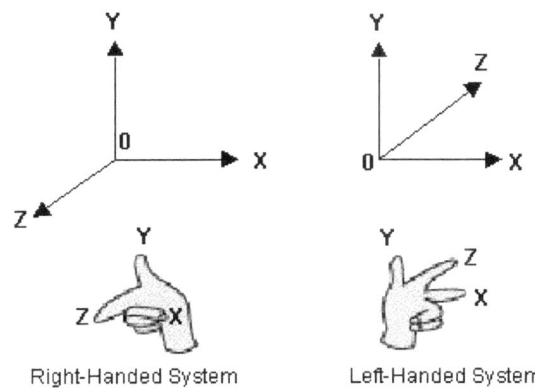

Right-Handed System   Left-Handed System

Linear combination of vectors: Let $\overline{a_1}, \overline{a_2}, \overline{a_3}, \ldots\ldots, \overline{a_n}$ be the n vectors and let $\bar{r}$ be any vector then $\bar{r} = x_1\overline{a_1} + x_2\overline{a_2} + x_3\overline{a_3} + \ldots\ldots + x_n\overline{a_n}$ for scalers $x_1, x_2, x_3, \ldots\ldots, x_n$ is called as a linear combination of the vectors.

Linearly dependent and Linearly independent vectors:

(i) Let $\overline{a_1}, \overline{a_2}, \overline{a_3}, \ldots\ldots, \overline{a_n}$ be the n vectors and $x_1, x_2, x_3, \ldots\ldots, x_n$ be the n scalars. If the linear combination $x_1\overline{a_1} + x_2\overline{a_2} + x_3\overline{a_3} + \ldots\ldots + x_n\overline{a_n} = \bar{0}$ where not all the scalars are zero then the vectors are called as linearly dependent vectors.

(ii) Let $\overline{a_1}, \overline{a_2}, \overline{a_3}, \ldots\ldots, \overline{a_n}$ be the n vectors and $x_1, x_2, x_3, \ldots\ldots, x_n$ be the n scalars. If the linear combination $x_1\overline{a_1} + x_2\overline{a_2} + x_3\overline{a_3} + \ldots\ldots + x_n\overline{a_n} = \bar{0}$ where all the scalars are zero then the vectors are called as linearly independent vectors.

Angle between two vectors: Let $\overline{OA} = \bar{a}$ and $\overline{OB} = \bar{b}$ be two non-zero vectors and θ be the angle with $0° \leq \theta \leq 180°$. Then the angle is written as $(\bar{a}, \bar{b})$

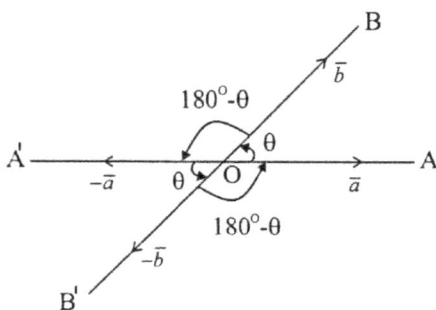

Properties of angle: Let $(\bar{a}, \bar{b})$ be the angle between the vectors $\bar{a}$ and $\bar{b}$. Then

(i) $(\bar{a}, \bar{b}) = (\bar{b}, \bar{a}) = (-\bar{a}, -\bar{b}) = (-\bar{b}, -\bar{a})$

(ii) $(\bar{a}, -\bar{b}) = (-\bar{a}, \bar{b}) = 180° - (\bar{a}, \bar{b})$

(iii) $(m\bar{a}, n\bar{b}) = (\bar{a}, \bar{b}) = (-m\bar{a}, -n\bar{b})$

(iv) $(m\bar{a}, -n\bar{b}) = (-m\bar{a}, n\bar{b}) = 180° - (\bar{a}, \bar{b})$

(v) $(\bar{a}, \bar{b}) = 0 \Leftrightarrow \bar{a}, \bar{b}$ are like vectors

(vi) $(\bar{a}, \bar{b}) = 90° \Leftrightarrow \bar{a}, \bar{b}$ are perpendicular vectors

(vii) $(\bar{a}, \bar{b}) = 180° \Leftrightarrow \bar{a}, \bar{b}$ are unlike vectors

Section formula:

(i) Let C be the point divides the line segment joining the points A and B with position vectors $\bar{a}, \bar{b}$ respectively in the ratio $m:n$ internally then the position vector of C is $\overline{OC} = \dfrac{m\bar{b}+n\bar{a}}{m+n}$

(ii) Let D be the point divides the line segment joining the points A and B with position vectors $\bar{a}, \bar{b}$ respectively in the ratio $m:n$ externally then the position vector of D is $\overline{OD} = \dfrac{m\bar{b}-n\bar{a}}{m-n}$

Midpoint: Let C be the midpoint of the line segment joining the points A and B with position vectors $\bar{a}, \bar{b}$ respectively then the position vector of C is

$$\overline{OC} = \dfrac{\bar{a}+\bar{b}}{2}$$

Trisection points: The points which divides the line segment joining two points in the ratio 1:2 or 2:1 internally are called as trisection points.

Centroid of a triangle: Let $\bar{a}, \bar{b}, \bar{c}$ be the position vectors of vertices of a triangle ABC then the position vector of centroid is given by

$$\overline{OG} = \dfrac{\bar{a}+\bar{b}+\bar{c}}{3}$$

Centroid of a tetrahedron: Let $\bar{a}, \bar{b}, \bar{c}, \bar{d}$ be the position vectors of vertices of a tetrahedron ABCD then the position vector of centroid is given by

$$\overline{OG} = \frac{\bar{a} + \bar{b} + \bar{c} + \bar{d}}{4}$$

Collinear vectors: The vectors lie on a same line or on parallel lines are called as collinear vectors.

(i) If two vectors $\bar{a}, \bar{b}$ are collinear then we can write $\bar{a} = \lambda \bar{b}$ for some real number λ

(ii) The two vectors $\bar{a} = a_1 \bar{\imath} + b_1 \bar{\jmath} + c_1 \bar{k}$ and $\bar{b} = a_2 \bar{\imath} + b_2 \bar{\jmath} + c_2 \bar{k}$ are collinear when $\dfrac{a_1}{a_2} = \dfrac{b_1}{b_2} = \dfrac{c_1}{c_2}$

Coplanar vectors: Three are more vectors are said to be coplanar vectors when they lie on a same plane or on parallel planes.

To show the three vectors $\bar{a} = a_1 \bar{\imath} + b_1 \bar{\jmath} + c_1 \bar{k}$; $\bar{b} = a_2 \bar{\imath} + b_2 \bar{\jmath} + c_2 \bar{k}$ and $\bar{c} = a_3 \bar{\imath} + b_3 \bar{\jmath} + c_3 \bar{k}$ to be coplanar we have to show that

$$\begin{vmatrix} a_1 & b_1 & c_1 \\ a_2 & b_2 & c_2 \\ a_3 & b_3 & c_3 \end{vmatrix} = 0$$

Vector equations of lines:

(i) The vector equation of a line passing through the point $A(\bar{a})$ and parallel to the vector $\bar{b}$ is $\bar{r} = \bar{a} + t\bar{b} \ \forall t \in \mathbb{R}$

(ii) The vector equation of the line passing through the points $A(\bar{a})$ and $B(\bar{b})$ is

$$\bar{r} = (1 - t)\bar{a} + t\bar{b} \ \forall t \in \mathbb{R}$$

Vector equations of planes:

(i) The vector equation of the plane passing through the point $(\bar{a})$ and parallel to the vectors $\bar{b}$ and $\bar{c}$ is $\bar{r} = \bar{a} + t\bar{b} + s\bar{c} \ \forall t, s \in \mathbb{R}$

(ii) The vector equation of the plane passing through the points $A(\bar{a})$ and $B(\bar{b})$ and parallel to the vector $\bar{c}$ is $\bar{r} = (1 - t)\bar{a} + t\bar{b} + s\bar{c} \ \forall t, s \in \mathbb{R}$

(iii) The vector equation of the plane passing through the points $A(\bar{a})$; $B(\bar{b})$ and $C(\bar{c})$ is $\bar{r} = (1 - t - s)\bar{a} + t\bar{b} + s\bar{c} \ \forall t, s \in \mathbb{R}$

Some important points related to triangle, parallelogram, hexagon:

(i) If AD is the internal angular bisector of A of triangle ABC then D divides BC in the ratio AB:AC. Then the vector of angular bisector is given by

$$\overline{AD} = \frac{|\overline{AC}|\overline{AB} + |\overline{AB}|\overline{AC}}{|\overline{AB}| + |\overline{AC}|}$$

(ii) Let $BC = a\,;\,CA = b\,;\,AB = c$ in a triangle ABC then the position vector of incenter is given by $\dfrac{a\bar{a}+b\bar{b}+c\bar{c}}{a+b+c}$

(iii) Let S, O be the circumcenter and orthocenter of a triangle ABC then

    (a) $\overline{SA} + \overline{SB} + \overline{SC} = \overline{SO}$

    (b) $\overline{OA} + \overline{OB} + \overline{OC} = 2\overline{OS}$

(iv) Let ABCD be a parallelogram. Then the diagonals $\overline{AC} = \overline{AB} + \overline{BC}$ and $\overline{BD} = \overline{BC} - \overline{AB}$

(v) Let ABCDEF be a regular hexagon with center O. Then
$\overline{AB} + \overline{BC} + \overline{CD} + \overline{DE} + \overline{EF} + \overline{FA} = 3\overline{AD} = 6\overline{AO}$

Position of a point: Let the position vectors of A and B be $\bar{a}, \bar{b}$ respectively then the point $\overline{OC} = p\bar{a} + q\bar{b}$ lies

    (i) inside of $\triangle OAB$, if $p > 0, q > 0$ and $p + q < 1$

    (ii) outside of $\triangle OAB$, and inside of $\angle AOB$ if $p > 0, q > 0$ and $p + q > 1$

    (iii) outside of $\triangle OAB$, and inside of $\angle OAB$ if $p < 0, q > 0$ and $p + q < 1$

    (iv) outside of $\triangle OAB$, and inside of $\angle OBA$ if $p > 0, q < 0$ and $p + q < 1$

Rotation of a vector about an axis: Let $\bar{a} = (a_1, a_2, a_3)$ be a vector in right-handed system. If the system rotated through an angle α about

    (i) X-axis then the new components are

    $(a_1, a_2\cos\alpha + a_3\sin\alpha, -a_2\sin\alpha + a_3\cos\alpha)$

    (ii) Y-axis then the new components are

    $(-a_3\sin\alpha + a_1\cos\alpha, a_2, a_3\cos\alpha + a_1\sin\alpha)$

    (iii) Z-axis then the new components are

    $(a_1\cos\alpha + a_2\sin\alpha, -a_1\sin\alpha + a_2\cos\alpha, a_3)$

44. MULTIPLICATION OF VECTORS

DOT PRODUCT (OR) SCALAR PRODUCT

Dot product or Scalar product: Let $\bar{a}, \bar{b}$ be two non-zero vectors. Then the dot product is denoted as $\bar{a} \cdot \bar{b}$ and is defined as $\bar{a} \cdot \bar{b} = |\bar{a}||\bar{b}| Cos\theta$ where θ is the angle between $\bar{a}, \bar{b}$

$$Cos\theta = \frac{\bar{a} \cdot \bar{b}}{|\bar{a}||\bar{b}|}$$

(i) If $\bar{a} \cdot \bar{b} = 0$ then $(\bar{a}, \bar{b}) = 90^0$ i.e. the vectors are perpendicular

(ii) If $\bar{a} \cdot \bar{b} < 0$ then the angle is acute

(iii) $\bar{a} \cdot \bar{b} > 0$ then the angle is obtuse

Dot product between unit vectors: Let $\bar{\imath}, \bar{\jmath}, \bar{k}$ be the unit vectors acting along the positive directions of X, Y, Z-axes respectively. Then

(i) $\bar{\imath} \cdot \bar{\imath} = \bar{\jmath} \cdot \bar{\jmath} = \bar{k} \cdot \bar{k} = 1$ **(ii)** $\bar{\imath} \cdot \bar{\jmath} = \bar{\jmath} \cdot \bar{k} = \bar{k} \cdot \bar{\imath} = 0$

Evaluation of Dot product: Let $\bar{a} = a_1\bar{\imath} + b_1\bar{\jmath} + c_1\bar{k}$; $\bar{b} = a_2\bar{\imath} + b_2\bar{\jmath} + c_2\bar{k}$ be two vectors then $\bar{a} \cdot \bar{b} = a_1 a_2 + b_1 b_2 + c_1 c_2$

Properties of dot product:

(i) $\bar{a} \cdot \bar{b} = \bar{b} \cdot \bar{a}$ **(ii)** $\bar{a} \cdot \bar{a} = |\bar{a}|^2$ **(iii)** $\bar{a} \cdot (\bar{b} + \bar{c}) = \bar{a} \cdot \bar{b} + \bar{a} \cdot \bar{c}$

(iv) $(l\bar{a}) \cdot (m\bar{b}) = lm(\bar{a} \cdot \bar{b}) = (m\bar{a}) \cdot (l\bar{b})$ where l, m are scalars

(v) $|\bar{a} + \bar{b}|^2 = |\bar{a}|^2 + |\bar{b}|^2 + 2(\bar{a} \cdot \bar{b})$

(vi) $|\bar{a} - \bar{b}|^2 = |\bar{a}|^2 + |\bar{b}|^2 - 2(\bar{a} \cdot \bar{b})$

(vii) $|\bar{a} + \bar{b} + \bar{c}|^2 = |\bar{a}|^2 + |\bar{b}|^2 + |\bar{c}|^2 + 2(\bar{a} \cdot \bar{b} + \bar{b} \cdot \bar{c} + \bar{c} \cdot \bar{a})$

Vector component or orthogonal projection:

(i) The vector component or orthogonal projection of $\bar{a}$ on $\bar{b}$ is given by

$$\frac{(\bar{a} \cdot \bar{b})\bar{b}}{|\bar{b}|^2}$$

(ii) The vector component or orthogonal projection of $\bar{a}$ perpendicular to $\bar{b}$ is given by $\bar{a} - \frac{(\bar{a} \cdot \bar{b})\bar{b}}{|\bar{b}|^2}$

Length of the projection:

(i) The length of the projection $\bar{a}$ on $\bar{b}$ is given by $\dfrac{|\bar{a}\cdot\bar{b}|}{|\bar{a}|}$

(ii) The length of the projection $\bar{b}$ on $\bar{a}$ is given by $\dfrac{|\bar{a}\cdot\bar{b}|}{|\bar{b}|}$

Equation of plane:

(i) The vector equation of the plane which is at a distance of p units from the origin and $\hat{n}$ be the unit vector perpendicular to the plane is $\bar{r}\cdot\hat{n} = p$

(ii) The vector equation of the plane passing through the point $A(\bar{a})$ and perpendicular the vector $\hat{n}$ is $(\bar{r}-\bar{a})\cdot\hat{n} = 0$

Angle between a line and a plane:
The angle between a line and a plane is the complement of the angle between the line and normal to the plane.

If θ is the angle between the line $\bar{r} = \bar{a} + t\bar{b}$ and the plane $\bar{r}\cdot\bar{m} = p$ then θ is given by

$$Cos(90^0 - \theta) = Sin\theta = \dfrac{\bar{b}\cdot\bar{m}}{|\bar{b}||\bar{m}|}$$

Angle between two planes:
If θ is the angle between the planes

$\bar{r}\cdot\widehat{n_1} = p_1$ and $\bar{r}\cdot\widehat{n_2} = p_2$ then $Cos\theta = \dfrac{\widehat{n_1}\cdot\widehat{n_2}}{|\widehat{n_1}||\widehat{n_2}|}$

Perpendicular distance:

The perpendicular distance from the origin to the plane $(\bar{r}-\bar{a})\cdot\hat{n} = 0$ is

$\dfrac{|\bar{a}\cdot\hat{n}|}{|\hat{n}|}$

Work done:

(i) If $\bar{F}$ is the force and $\bar{s}$ is the displacement of a particle then the work done by the particle is $\bar{F}\cdot\bar{s}$

(ii) If $\bar{F}$ is the force acting on a particle displaces from A to B then work done is $\bar{F}\cdot\overline{AB}$

(iii) If $\bar{F}$ is the resultant of the forces $\bar{F}_1, \bar{F}_2, \bar{F}_3, \ldots \ldots \bar{F}_n$ acting on a particle displaces from A to B then work done is

$\bar{F}\cdot\overline{AB} = (\bar{F}_1 + \bar{F}_2 + \bar{F}_3, + \cdots \ldots + \bar{F}_n)\cdot\overline{AB}$

CROSS PRODUCT (OR) VECTOR PRODUCT

Cross product or vector product: Let $\bar{a}, \bar{b}$ be two non-zero vectors. Then the cross product is denoted as $\bar{a} \times \bar{b}$ and is defined as $\bar{a} \times \bar{b} = |\bar{a}||\bar{b}|Sin\theta \ \hat{n}$ where θ is the angle between $\bar{a}, \bar{b}$ and $\hat{n}$ is the unit vector perpendicular to both $\bar{a}$ and $\bar{b}$

$$Sin\theta = \frac{|\bar{a} \times \bar{b}|}{|\bar{a}||\bar{b}|}$$

If $\bar{a} \times \bar{b} = \bar{0}$ then either $\bar{a} = \bar{0}$ or $\bar{b} = \bar{0}$ or $\bar{a}$ and $\bar{b}$ are parallel vectors

Cross product between unit vectors: Let $\bar{i}, \bar{j}, \bar{k}$ be the unit vectors acting along the positive directions of X, Y, Z-axes respectively. Then

(i) $\bar{i} \times \bar{i} = \bar{j} \times \bar{j} = \bar{k} \times \bar{k} = \bar{0}$ (ii) $\bar{i} \times \bar{j} = \bar{k}$; $\bar{j} \times \bar{k} = \bar{i}$; $\bar{k} \times \bar{i} = \bar{j}$

(iii) $\bar{j} \times \bar{i} = -\bar{k}$; $\bar{k} \times \bar{j} = -\bar{i}$; $\bar{i} \times \bar{k} = -\bar{j}$

Evaluation of cross product: Let $\bar{a} = a_1\bar{i} + b_1\bar{j} + c_1\bar{k}$; $\bar{b} = a_2\bar{i} + b_2\bar{j} + c_2\bar{k}$ be two vectors then

$$\bar{a} \times \bar{b} = \begin{vmatrix} \bar{i} & \bar{j} & \bar{k} \\ a_1 & b_1 & c_1 \\ a_2 & b_2 & c_2 \end{vmatrix}$$

Properties of cross product:

(i) $\bar{a} \times \bar{b} \neq \bar{b} \times \bar{a}$ (ii) $\bar{a} \times \bar{b} = -(\bar{b} \times \bar{a})$ (iii) $|\bar{a} \times \bar{b}| = |\bar{b} \times \bar{a}|$

(iv) $\bar{a} \times \bar{a} = \bar{0}$ (v) $\bar{a} \times (\bar{b} + \bar{c}) = \bar{a} \times \bar{b} + \bar{a} \times \bar{c}$

(vi) $-\bar{a} \times \bar{b} = \bar{a} \times -(\bar{b}) = -(\bar{a} \times \bar{b})$

(iv) $(l\bar{a}) \times (m\bar{b}) = lm(\bar{a} \times \bar{b}) = (m\bar{a}) \times (l\bar{b})$ where l, m are scalars

Perpendicular vectors and Unit perpendicular vectors:

(i) The vector perpendicular to both $\bar{a}$ and $\bar{b}$ is $\lambda(\bar{a} \times \bar{b})$ for $\lambda \in \mathbb{R}$

(ii) The unit vector perpendicular to both $\bar{a}$ and $\bar{b}$ is $\pm \frac{(\bar{a} \times \bar{b})}{|\bar{a} \times \bar{b}|}$

(iii) The vector perpendicular to both $\bar{a}$ and $\bar{b}$ having magnitude k units is

$$\pm k \frac{(\bar{a} \times \bar{b})}{|\bar{a} \times \bar{b}|}$$

(iv)   The unit vector perpendicular to the plane $\overline{ABC}$ is $\pm \frac{(\overline{AB} \times \overline{AC})}{|\overline{AB} \times \overline{AC}|}$

Perpendicular distance:

(i) The perpendicular distance from a point P to the line joining the points A and B is $\frac{|\overline{AP} \times \overline{AB}|}{|\overline{AB}|}$

(ii) The length of the projection of $\bar{b}$ on a vector perpendicular to $\bar{a}$ in the plane generated by $\bar{a} \cdot \bar{b}$ is $\frac{|\bar{a} \times \bar{b}|}{|\bar{a}|}$

Vector equation of line:

(i) The vector equation of the line passing through the point $A(\bar{a})$ and parallel to the vector $\bar{b}$ is $(\bar{r} - \bar{a}) \times \bar{b} = \bar{0}$

(ii) The vector equation of the line passing through the points $A(\bar{a})$ and $B(\bar{b})$ is $(\bar{r} - \bar{a}) \times (\bar{b} - \bar{a}) = \bar{0}$

Vector areas and areas:

(i) The vector area of the triangle ABC $= \frac{1}{2}(\overline{AB} \times \overline{AC}) = \frac{1}{2}(\overline{BC} \times \overline{BA}) = \frac{1}{2}(\overline{CA} \times \overline{CB})$

(ii) The area of the triangle ABC $= \frac{1}{2}|\overline{AB} \times \overline{AC}| = \frac{1}{2}|\overline{BC} \times \overline{BA}| = \frac{1}{2}|\overline{CA} \times \overline{CB}|$

(iii) If $\bar{a}$ and $\bar{b}$ are the adjacent sides of a triangle then the vector area of the triangle is $\frac{1}{2}(\bar{a} \times \bar{b})$

(iv) If $\bar{a}$ and $\bar{b}$ are the adjacent sides of a triangle then the area of the triangle is $\frac{1}{2}|\bar{a} \times \bar{b}|$

(v) If $\bar{a}$ and $\bar{b}$ are the adjacent sides of a parallelogram then the vector area of the parallelogram is $(\bar{a} \times \bar{b})$

(vi) If $\bar{a}$ and $\bar{b}$ are the adjacent sides of a parallelogram then the area of the parallelogram is $|\bar{a} \times \bar{b}|$

(vii) If $\bar{a}$ and $\bar{b}$ are the diagonals of a parallelogram then the vector area of the parallelogram is $\frac{1}{2}(\bar{a} \times \bar{b})$

(vi) If $\bar{a}$ and $\bar{b}$ are the diagonals of a parallelogram then the area of the parallelogram is $\frac{1}{2}|\bar{a} \times \bar{b}|$

Torque or Moment of force: Let O be the point of reference and $\overline{OP} = \bar{r}$ be the position vector of P on the line of action of a force $\bar{F}$ then the moment of force is $\bar{r} \times \bar{F}$

SCALAR TRIPLE PRODUCT

Scalar triple product: The scalar triple product between three vectors $\bar{a}, \bar{b}, \bar{c}$ is denoted by $(\bar{a} \times \bar{b}) \cdot \bar{c}$ or $\bar{a} \cdot (\bar{b} \times \bar{c})$

The scalar triple product is also denoted as $[\bar{a} \ \ \bar{b} \ \ \bar{c}]$

Properties of scalar triple product:

(i) $[\bar{a} \ \ \bar{b} \ \ \bar{c}] = [\bar{b} \ \ \bar{c} \ \ \bar{a}] = [\bar{c} \ \ \bar{a} \ \ \bar{b}]$
$= -[\bar{b} \ \ \bar{a} \ \ \bar{c}] = -[\bar{a} \ \ \bar{c} \ \ \bar{b}] = -[\bar{c} \ \ \bar{b} \ \ \bar{a}]$

(ii) $[\bar{i} \ \ \bar{j} \ \ \bar{k}] = [\bar{j} \ \ \bar{k} \ \ \bar{i}] = [\bar{k} \ \ \bar{i} \ \ \bar{j}] = 1$

(iii) If $\bar{a} = a_1\bar{i} + b_1\bar{j} + c_1\bar{k}$; $\bar{b} = a_2\bar{i} + b_2\bar{j} + c_2\bar{k}$ and $\bar{c} = a_3\bar{i} + b_3\bar{j} + c_3\bar{k}$

then $[\bar{a} \ \ \bar{b} \ \ \bar{c}] = \begin{vmatrix} a_1 & b_1 & c_1 \\ a_2 & b_2 & c_2 \\ a_3 & b_3 & c_3 \end{vmatrix}$

Coplanar vectors:

(i) The three vectors $\bar{a}, \bar{b}, \bar{c}$ are coplanar when $[\bar{a} \ \ \bar{b} \ \ \bar{c}] = 0$

(ii) The four points A, B, C, D are coplanar when $[\overline{AB} \ \ \overline{AC} \ \ \overline{AD}] = 0$

Vector equation of plane:

(i) The vector equation of plane passing through three points $A(\bar{a})$; $B(\bar{b})$ and $C(\bar{c})$ is $[\bar{r} - \bar{a} \ \ \bar{b} - \bar{a} \ \ \bar{c} - \bar{a}] = 0$ or $[\bar{r} \ \ \bar{b} \ \ \bar{c}] + [\bar{r} \ \ \bar{c} \ \ \bar{a}] + [\bar{r} \ \ \bar{a} \ \ \bar{b}] = [\bar{a} \ \ \bar{b} \ \ \bar{c}]$

(ii) The vector equation passes through the points $A(\bar{a})$; $B(\bar{b})$ and parallel to the vector $\bar{c}$ is $[\bar{r} - \bar{a} \ \ \bar{b} - \bar{a} \ \ \bar{c}] = 0$

(iii) The vector equation passes through the point $A(\bar{a})$ and parallel to the vectors $\bar{b}$ and $\bar{c}$ is $[\bar{r} - \bar{a} \ \ \bar{b} \ \ \bar{c}] = 0$

Perpendicular vector and length of the perpendicular:

(i) The unit vector perpendicular to the plane containing three vectors $\bar{a}, \bar{b}, \bar{c}$ is $\pm \dfrac{(\bar{a} \times \bar{b}) + (\bar{b} \times \bar{c}) + (\bar{c} \times \bar{a})}{|(\bar{a} \times \bar{b}) + (\bar{b} \times \bar{c}) + (\bar{c} \times \bar{a})|}$

(ii) The length of the perpendicular from the origin to the plane containing three vectors $\bar{a}, \bar{b}, \bar{c}$ is $\dfrac{|[\bar{a} \ \ \bar{b} \ \ \bar{c}]|}{|(\bar{a} \times \bar{b}) + (\bar{b} \times \bar{c}) + (\bar{c} \times \bar{a})|}$

Distance between skew-lines: The shortest distance between the skew-lines $\bar{r} = \bar{a} + t\bar{b}$ and $\bar{r} = \bar{c} + s\bar{d}$ where t and s are scalars is $\dfrac{|[\bar{a} - \bar{c} \ \ \bar{b} \ \ \bar{d}]|}{|\bar{b} \times \bar{d}|}$

Volumes:

(i) The volume of the parallelopiped having coterminous edges $\bar{a}, \bar{b}, \bar{c}$ is $|[\bar{a} \quad \bar{b} \quad \bar{c}]|$

(ii) The volume of the parallelopiped having vertices A, B, C, D of coterminous edges is $|[\overline{AB} \quad \overline{AC} \quad \overline{AD}]|$

(iii) The volume of the tetrahedron having coterminous edges $\bar{a}, \bar{b}, \bar{c}$ is $\frac{1}{6}|[\bar{a} \quad \bar{b} \quad \bar{c}]|$

(iv) The volume of the tetrahedron having vertices A, B, C, D of coterminous edges is $\frac{1}{6}|[\overline{AB} \quad \overline{AC} \quad \overline{AD}]|$

(v) The volume of the triangular prism having coterminous edges $\bar{a}, \bar{b}, \bar{c}$ is $\frac{1}{2}|[\bar{a} \quad \bar{b} \quad \bar{c}]|$

(vi) The volume of the triangular prism having vertices A, B, C, D of coterminous edges is $\frac{1}{2}|[\overline{AB} \quad \overline{AC} \quad \overline{AD}]|$

Reciprocal system: If $\bar{a}, \bar{b}, \bar{c}$ are any three non-coplanar vectors then the vectors $\overline{a^1}, \overline{b^1}, \overline{c^1}$ defined by $\overline{a^1} = \frac{\bar{b} \times \bar{c}}{[\bar{a} \quad \bar{b} \quad \bar{c}]}; \overline{b^1} = \frac{\bar{c} \times \bar{a}}{[\bar{a} \quad \bar{b} \quad \bar{c}]}; \overline{c^1} = \frac{\bar{a} \times \bar{b}}{[\bar{a} \quad \bar{b} \quad \bar{c}]}$ is called as reciprocal system of $\bar{a}, \bar{b}, \bar{c}$

(i) $\bar{a} \cdot \overline{a^1} = \bar{b} \cdot \overline{b^1} = \bar{c} \cdot \overline{c^1} = 1$

(ii) $[\overline{a^1} \quad \overline{b^1} \quad \overline{c^1}] = [\bar{a} \quad \bar{b} \quad \bar{c}]$

Some important results:

(i) $[\bar{a} + \bar{b} \quad \bar{b} + \bar{c} \quad \bar{c} + \bar{a}] = 2[\bar{a} \quad \bar{b} \quad \bar{c}]$

(ii) $[\bar{a} \times \bar{b} \quad \bar{b} \times \bar{c} \quad \bar{c} \times \bar{a}] = [\bar{a} \quad \bar{b} \quad \bar{c}]^2$

(iii) $[\bar{a} - \bar{b} \quad \bar{b} - \bar{c} \quad \bar{c} - \bar{a}] = 0$

(iv) $[\bar{a} \quad \bar{b} \quad \bar{c}]^2 = \begin{vmatrix} \bar{a} \cdot \bar{a} & \bar{a} \cdot \bar{b} & \bar{a} \cdot \bar{c} \\ \bar{b} \cdot \bar{a} & \bar{b} \cdot \bar{b} & \bar{b} \cdot \bar{c} \\ \bar{c} \cdot \bar{a} & \bar{c} \cdot \bar{b} & \bar{c} \cdot \bar{c} \end{vmatrix}$

VECTOR TRIPLE PRODUCT

Vector triple product: The vector triple product between three vectors $\bar{a}, \bar{b}, \bar{c}$ is denoted by $(\bar{a} \times \bar{b}) \times \bar{c}$ or $\bar{a} \times (\bar{b} \times \bar{c})$ and is defined as $(\bar{a} \times \bar{b}) \times \bar{c} = (\bar{a} \cdot \bar{c})\bar{b} - (\bar{b} \cdot \bar{c})\bar{a}$ and $\bar{a} \times (\bar{b} \times \bar{c}) = (\bar{a} \cdot \bar{c})\bar{b} - (\bar{a} \cdot \bar{b})\bar{c}$

$(\bar{a} \times \bar{b}) \times \bar{c} \neq \bar{a} \times (\bar{b} \times \bar{c})$

SCALAR PRODUCT AMONG FOUR VECTORS

Scalar product among four vectors: The scalar product among the vectors $\bar{a}, \bar{b}, \bar{c}, \bar{d}$ is defined as $(\bar{a} \times \bar{b}) \cdot (\bar{c} \times \bar{d}) = (\bar{a} \cdot \bar{c})(\bar{b} \cdot \bar{d}) - (\bar{a} \cdot \bar{d})(\bar{b} \cdot \bar{c}) = \begin{vmatrix} \bar{a} \cdot \bar{c} & \bar{a} \cdot \bar{d} \\ \bar{b} \cdot \bar{c} & \bar{b} \cdot \bar{d} \end{vmatrix}$

$(\bar{a} \times \bar{b}) \cdot (\bar{a} \times \bar{b}) = |\bar{a} \times \bar{b}|^2 = |\bar{a}|^2|\bar{b}|^2 - |\bar{a} \cdot \bar{b}|^2$

VECTOR PRODUCT AMONG FOUR VECTORS

Vector product among four vectors: The vector product among the vectors $\bar{a}, \bar{b}, \bar{c}, \bar{d}$ is defined as $(\bar{a} \times \bar{b}) \times (\bar{c} \times \bar{d}) = [\bar{a} \ \bar{b} \ \bar{d}]\bar{c} - [\bar{a} \ \bar{b} \ \bar{c}]\bar{d}$

Linear combination: The linear combination of three vectors $\bar{a}, \bar{b}, \bar{c}$ is represented as

$$\bar{r} = \frac{[\bar{r} \ \bar{b} \ \bar{c}]}{[\bar{a} \ \bar{b} \ \bar{c}]}\bar{a} + \frac{[\bar{r} \ \bar{c} \ \bar{a}]}{[\bar{a} \ \bar{b} \ \bar{c}]}\bar{b} + \frac{[\bar{r} \ \bar{a} \ \bar{b}]}{[\bar{a} \ \bar{b} \ \bar{c}]}\bar{c}$$

CALCULUS

45. LIMITS, CONTINUITY AND DIFFERENTIABILITY

Neighbourhood of a real number: Let a be a real number and δ be a positive real number then $(a - \delta, a + \delta)$ is called as δ-neighbourhood of a.

$(a - \delta, a)$ is called the left δ-neighbourhood of a and $(a, a + \delta)$ is called the right δ-neighbourhood of a.

Deleted neighbourhood: $(a - \delta, a) \cup (a, a + \delta)$ is called the deleted neighbourhood of a.

Limit of a function: Let $f(x)$ be a function defined over a deleted neighbourhood of the real number a and l be a real number. If for every positive number ε (so small) then there exists a positive number δ such that $|f(x) - l| < \varepsilon$ for all x with $0 < |x - a| < \delta$.

Simply, we can say that $f(x)$ tends to l as x tends to a. Mathematically it shown that $\lim_{x \to a} f(x) = l$

Operations on limits: Let $\lim_{x \to a} f(x) = l$ and $\lim_{x \to a} g(x) = m$. Then

(i) $\lim_{x \to a}[f(x) + g(x)] = l + m$ **(ii)** $\lim_{x \to a}[f(x) - g(x)] = l - m$

(iii) $\lim_{x \to a}[f(x) \cdot g(x)] = lm$ **(iv)** $\lim_{x \to a}[f(x)/g(x)] = l/m$

(v) $\lim_{x \to a} k[f(x)] = kl$ **(vi)** $\lim_{x \to a}[f(x)]^n = l^n$

(vii) $\lim_{x \to a}[f(x)]^{g(x)} = l^m$ **(viii)** $\lim_{x \to a} f(g(x)) = f(m)$

Right handed limit: Let $f(x)$ be a function defined on $(a, a + \delta)$ and l be a real number. If for every positive number ε (so small) then there exists a positive number δ such that $|f(x) - l| < \varepsilon$ for all $x \in (a, a + \delta)$

i.e. $\lim_{x \to a^+} f(x) = l$

Left handed limit: Let $f(x)$ be a function defined on $(a - \delta, a)$ and l be a real number. If for every positive number ε (so small) then there exists a positive number δ such that $|f(x) - l| < \varepsilon$ for all $x \in (a - \delta, a)$ i.e. $\lim_{x \to a^-} f(x) = l$

Existence of Limit: $\lim_{x \to a} f(x) = l$ exists only when $\lim_{x \to a^+} f(x) = l = \lim_{x \to a^-} f(x)$

Infinite limits: Let $f(x)$ be a function defined over a deleted neighbourhood of the real number a. If for every positive number k (so large) then there exists a positive number δ such that $f(x) > k$ for all x with $0 < |x - a| < \delta$.

Simply, we can say that $f(x)$ tends to ∞ as x tends to a. Mathematically it shown that $\lim_{x \to a} f(x) = \infty$

Limit at infinity: Let $f(x)$ be a function defined over a deleted neighbourhood of the real number a and l be a real number. If for every positive number ε (so small) then there exists a positive number k (so large) such that $|f(x) - l| < \varepsilon$ for all x with $x > k$.

Simply, we can say that $f(x)$ tends to l as x tends to ∞. Mathematically it shown that $\lim\limits_{x \to \infty} f(x) = l$

Standard Limits:

(i) $\lim\limits_{x \to a} \dfrac{x^n - a^n}{x - a} = na^{n-1}$

(ii) $\lim\limits_{x \to a} \dfrac{x^m - a^m}{x^n - a^n} = \dfrac{m}{n} a^{m-n}$

(iii) $\lim\limits_{x \to 0} \dfrac{e^x - 1}{x} = 1$

(iv) $\lim\limits_{x \to 0} \dfrac{a^x - 1}{x} = \log_e a$

(v) $\lim\limits_{x \to 0} \dfrac{a^x - b^x}{x} = \log_e \left(\dfrac{a}{b}\right)$

(vi) $\lim\limits_{x \to 0} \dfrac{a^x - 1}{b^x - 1} = \log_b a$

(vii) $\lim\limits_{x \to 0} \dfrac{Sinx}{x} = 1$

(viii) $\lim\limits_{x \to 0} \dfrac{Tanx}{x} = 1$

(ix) $\lim\limits_{x \to 0} \dfrac{Sin(ax)}{x} = a$

(x) $\lim\limits_{x \to 0} \dfrac{Tan(ax)}{x} = a$

(xi) $\lim\limits_{x \to 0} \dfrac{Sin^{-1}x}{x} = 1$

(xii) $\lim\limits_{x \to 0} \dfrac{Tan^{-1}x}{x} = 1$

(xiii) $\lim\limits_{x \to 0} \dfrac{Sinx^0}{x} = \dfrac{\pi}{180}$

(xiv) $\lim\limits_{x \to 0} \dfrac{Tanx^0}{x} = \dfrac{\pi}{180}$

(xv) $\lim\limits_{x \to \infty} \dfrac{Sinx}{x} = 0$

(xvi) $\lim\limits_{x \to \infty} \dfrac{Cosx}{x} = 0$

(xvii) $\lim\limits_{x \to 0}(1 + x)^{1/x} = e$

(xviii) $\lim\limits_{x \to 0}(1 + ax)^{1/x} = e^a$

(xix) $\lim\limits_{x \to \infty}\left(1 + \dfrac{1}{x}\right)^x = e$

(xx) $\lim\limits_{x \to \infty}\left(1 + \dfrac{a}{x}\right)^x = e^a$

(xxi) $\lim\limits_{x \to 0} Sin\left(\dfrac{1}{x}\right) = $ Does not exist

(xxii) $\lim\limits_{x \to 0} Cos\left(\dfrac{1}{x}\right) = $ Does not exist

(xxiii) $\lim\limits_{x \to 0} x\, Sin\left(\dfrac{1}{x}\right) = 0$

(xxiv) $\lim\limits_{x \to 0} x\, Cos\left(\dfrac{1}{x}\right) = 0$

Indeterminate forms: $\dfrac{0}{0}, \dfrac{\infty}{\infty}, 0 \times \infty, 0^0, \infty^0, 1^\infty$ etc., are called indeterminate forms.

L.Hospital's Rule: If $\lim\limits_{x \to a} \dfrac{f(x)}{g(x)}$ is of the form $\dfrac{0}{0}, \dfrac{\infty}{\infty}$ then we use the L.Hospital's rule.

$\lim\limits_{x \to a} \dfrac{f(x)}{g(x)} = \lim\limits_{x \to a} \dfrac{f^1(x)}{g^1(x)}$

If $\lim\limits_{x \to a} \dfrac{f^1(x)}{g^1(x)}$ is of the form $\dfrac{0}{0}, \dfrac{\infty}{\infty}$ then again we use the L.Hospital's rule.

$\lim\limits_{x \to a} \dfrac{f^1(x)}{g^1(x)} = \lim\limits_{x \to a} \dfrac{f^{11}(x)}{g^{11}(x)}$

Continuing this process till we get determinate form.

Some more methods to evaluate indeterminate limits:

(i) If $\lim\limits_{x \to a} f(x) = 1$ and $\lim\limits_{x \to a} g(x) = \infty$. Then $\lim\limits_{x \to a}[f(x)]^{g(x)} = e^{\lim\limits_{x \to a} g(x)[f(x)-1]}$

(ii) If $\lim\limits_{x \to a} f(x) = 0$ and $\lim\limits_{x \to a} g(x) = \infty$. Then $\lim\limits_{x \to a}[f(x)]^{g(x)} = e^{\lim\limits_{x \to a} g(x) \log f(x)}$

(iii) $\lim\limits_{x \to 0} \left(\dfrac{a_1{}^x + a_2{}^x + a_3{}^x + \cdots\cdots + a_n{}^x}{n} \right)^{1/x} = (a_1. a_2. a_3 \ldots \ldots a_n)^{1/n}$

(iv) $\lim\limits_{x \to \infty} \left(\dfrac{a_1{}^{1/x} + a_2{}^{1/x} + a_3{}^{1/x} + \cdots\cdots + a_n{}^{1/x}}{n} \right)^{x} = (a_1. a_2. a_3 \ldots \ldots a_n)^{1/n}$

Sandwich theorem or Squeeze Principle: Let $f(x), g(x), h(x)$ are functions such that $f(x) \leq g(x) \leq h(x)$ then $\lim\limits_{x \to a} f(x) \leq \lim\limits_{x \to a} g(x) \leq \lim\limits_{x \to a} h(x)$ and also

If $\lim\limits_{x \to a} f(x) = \lim\limits_{x \to a} h(x) = l$ then $\lim\limits_{x \to a} g(x) = l$

Some expansions useful to evaluate limits:

(i) $e^x = 1 + \dfrac{x}{1!} + \dfrac{x^2}{2!} + \dfrac{x^3}{3!} + \ldots\ldots\ldots$

(ii) $e^{-x} = 1 - \dfrac{x}{1!} + \dfrac{x^2}{2!} - \dfrac{x^3}{3!} + \ldots\ldots\ldots$

(iii) $a^x = 1 + \dfrac{x}{1!}(\log_e a) + \dfrac{x^2}{2!}(\log_e a)^2 + \dfrac{x^3}{3!}(\log_e a)^3 + \ldots\ldots\ldots$

(iv) $a^{-x} = 1 - \dfrac{x}{1!}(\log_e a) + \dfrac{x^2}{2!}(\log_e a)^2 - \dfrac{x^3}{3!}(\log_e a)^3 + \ldots\ldots\ldots$

(v) $\log_e(1 + x) = x - \dfrac{x^2}{2} + \dfrac{x^3}{3} - \dfrac{x^4}{4} + \ldots\ldots\ldots$ for $|x| < 1$

(vi) $\log_e(1 - x) = -\left(x + \dfrac{x^2}{2} + \dfrac{x^3}{3} + \dfrac{x^4}{4} + \ldots\ldots\ldots \right)$ for $|x| < 1$

(vii) $Sinx = x - \dfrac{x^3}{3!} + \dfrac{x^5}{5!} - \ldots\ldots\ldots$

(viii) $Cosx = 1 - \dfrac{x^2}{2!} + \dfrac{x^4}{4!} - \cdots\ldots\ldots$

(ix) $Tanx = x + \dfrac{x^3}{3} + \dfrac{2x^5}{15} + \cdots\ldots\ldots$

CONTINUITY

Continuous at a point: Let $f(x)$ be a function defined over a deleted neighbourhood of the real number a then f is said to be continuous at a when both $\lim_{x \to a^+} f(x)$ and $\lim_{x \to a^-} f(x)$ exist and both are equal to $f(a)$ i.e. $\lim_{x \to a^+} f(x) = \lim_{x \to a^-} f(x) = f(a)$

Continuous in open interval: The function f is said to be continuous in an open interval (a, b) when f is continuous at each and every point in the interval (a, b)

Continuous in closed interval: The function f is said to be continuous in a closed interval $[a, b]$ when

(i) f is continuous at each and every point in the interval (a, b)

(ii) $\lim_{x \to a^+} f(x) = f(a)$

(iii) $\lim_{x \to b^-} f(x) = f(b)$

Single point continuity: Functions which are continuous only at one point are said to be have single point continuity.

Properties: If f and g are continuous functions at $x = a$ then the following are continuous at $x = a$

(i) $f + g$ (ii) $f - g$ (iii) $f \cdot g$ (iv) f/g (v) kf

Discontinuous: The function f is said to be discontinuous at $x = a$ when it is not continuous at $x = a$

Types of discontinuity:

(i) Discontinuity of first kind or Removable discontinuity:

If $\lim_{x \to a} f(x)$ exist but it is not equal to $f(a)$ or $f(a)$ is not defined then f is said to have a removable discontinuity at $x = a$

Removable discontinuity is of two types

(a) Missing point discontinuity: If $\lim_{x \to a} f(x)$ exist and $f(a)$ is not defined then the discontinuity is called as missing point discontinuity

(b) Isolated point discontinuity: : If $\lim_{x \to a} f(x)$ exist and $f(a)$ is defined but $\lim_{x \to a} f(x) \neq f(a)$ then the discontinuity is called as isolated point discontinuity

(ii) Discontinuity of second kind or Irremovable discontinuity:

If $\lim_{x \to a} f(x)$ does not exist then f is said to have a irremovable discontinuity at $x = a$

Irremovable discontinuity is of three types

(a) Finite discontinuity or Jumping discontinuity:

If $\lim_{x \to a^+} f(x)$ and $\lim_{x \to a^-} f(x)$ both are exist but not equal then the discontinuity is called as jumping discontinuity.

(b) Infinite discontinuity: If at least one of the $\lim_{x \to a^+} f(x)$ and $\lim_{x \to a^-} f(x)$ is $\pm\infty$ then it is called infinite discontinuity.

(c) Oscillatory discontinuity: If the limit oscillates between two finite quantities then it is called as oscillatory discontinuity.

Intermediate value theorem: Let $f(x)$ be continuous on an interval I and a,b are any points in the interval. If y_0 is a number between $f(a)$ and $f(b)$ then there exists a number c between a and b such that $f(c) = y_0$

DIFFERENTIABILITY

Differentiable function: Let $f(x)$ be a real valued function then f is said to be differentiable when $\lim_{h \to 0} \frac{f(x+h) - f(x)}{h}$ exists finitely. It is denoted by $f^1(x)$ or $\frac{dy}{dx}$

Differentiability at a point: Let $f(x)$ be a real valued function then f is said to be differentiable at a point x=a when is $f^1(a)$ exist finitely where

$$f^1(a) = \lim_{h \to 0} \frac{f(a+h) - f(a)}{h} = \lim_{x \to a} \frac{f(x) - f(a)}{x - a}$$

Right hand derivative: The right hand derivative of $f(x)$ at x=a is denoted as

$$f^1(a^+) = \lim_{h \to 0} \frac{f(a+h) - f(a)}{h}$$

Left hand derivative: The left hand derivative of $f(x)$ at x=a is denoted as

$$f^1(a^-) = \lim_{h \to 0} \frac{f(a-h) - f(a)}{-h}$$

The function $f(x)$ is differentiable only when $\lim_{h \to 0} \frac{f(a+h) - f(a)}{h} = \lim_{h \to 0} \frac{f(a-h) - f(a)}{-h}$

i.e. LHD=RHD

Differentiability in open interval: The function f is said to be differentiable in an open interval (a, b) when f is differentiable at each and every point in the interval (a, b)

Differentiability in closed interval: The function f is said to be differentiable in a closed interval $[a, b]$ when

(i) f is differentiable at each and every point in the interval (a, b)

(ii) f is differentiable from the right at a

(iii) f is differentiable from the left at b

Key points:

(i) If f is differentiable then f is continuous

(ii) If f is not continuous then f is not differentiable

(iii) If f, g are differentiable then $f \pm g$ and $f.g$ are differentiable

(iv) If f is differentiable and g is not differentiable then $f \pm g$ are not differentiable and $f.g$ is may or may not differentiable

(v) If f, g are not differentiable then $f \pm g$ and $f.g$ are may or may not differentiable

46. DIFFERENTIATION

First principle of derivative: Let $y = f(x)$ be a function. Then the derivative of f is defined as

$$f'(x) \text{ or } \frac{dy}{dx} = \lim_{h \to 0} \frac{f(x+h) - f(x)}{h}$$

Derivatives of standard functions:

(i) $\frac{d}{dx}(Constant) = 0$

(ii) $\frac{d}{dx}(e^x) = e^x$

(iii) $\frac{d}{dx}(a^x) = a^x \log_e a$

(iv) $\frac{d}{dx}(x^n) = nx^{n-1}$

(v) $\frac{d}{dx}(\sqrt{x}) = \frac{1}{2\sqrt{x}}$

(vi) $\frac{d}{dx}\left(\frac{1}{x}\right) = \frac{-1}{x^2}$

(vii) $\frac{d}{dx}(|x|) = \frac{|x|}{x} \ (x \neq 0)$

(viii) $\frac{d}{dx}(\log_e |x|) = \frac{1}{x}$

(ix) $\frac{d}{dx}(Sinx) = Cosx$

(x) $\frac{d}{dx}(Cosx) = -Sinx$

(xi) $\frac{d}{dx}(Tanx) = Sec^2 x$

(xii) $\frac{d}{dx}(Cotx) = -Cosec^2 x$

(xiii) $\frac{d}{dx}(Secx) = Secx.Tanx$

(xiv) $\frac{d}{dx}(Cosecx) = -Cosecx.Cotx$

(xv) $\frac{d}{dx}(Sin^{-1}x) = \frac{1}{\sqrt{1-x^2}} \ \forall x \in (-1,1)$

(xvi) $\frac{d}{dx}(Cos^{-1}x) = \frac{-1}{\sqrt{1-x^2}} \ \forall x \in (-1,1)$

(xvii) $\frac{d}{dx}(Tan^{-1}x) = \frac{1}{1+x^2} \ \forall x \in \mathbb{R}$

(xviii) $\frac{d}{dx}(Cot^{-1}x) = \frac{-1}{1+x^2} \ \forall x \in \mathbb{R}$

(xix) $\frac{d}{dx}(Sec^{-1}x) = \frac{1}{|x|\sqrt{x^2-1}} \ \forall x \in \mathbb{R} - [-1,1]$

(xx) $\frac{d}{dx}(Cosec^{-1}x) = \frac{-1}{|x|\sqrt{x^2-1}} \ \forall x \in \mathbb{R} - [-1,1]$

(xxi) $\frac{d}{dx}(Sinhx) = Coshx$

(xxii) $\frac{d}{dx}(Coshx) = Sinhx$

(xxiii) $\frac{d}{dx}(Tanhx) = Sech^2 x$

(xxiv) $\frac{d}{dx}(Cothx) = -Cosech^2 x$

(xxv) $\frac{d}{dx}(Sechx) = -Sechx.Tanhx$

(xxvi) $\frac{d}{dx}(Cosechx) = -Cosechx.Cothx$

(xxvii) $\frac{d}{dx}(Sinh^{-1}x) = \frac{1}{\sqrt{1+x^2}}$

(xxviii) $\frac{d}{dx}(Cosh^{-1}x) = \frac{1}{\sqrt{x^2-1}} \ \forall x \notin (-1,1)$

(xxiv) $\frac{d}{dx}(Tanh^{-1}x) = \frac{1}{1-x^2} \ \forall x \in (-1,1)$

(xxv) $\frac{d}{dx}(Coth^{-1}x) = \frac{1}{1-x^2}$ $\forall x \in \mathbb{R} - [-1,1]$

(xxvi) $\frac{d}{dx}(Sech^{-1}x) = \frac{-1}{|x|\sqrt{1-x^2}}$ $\forall x \in (0,1)$

(xx) $\frac{d}{dx}(Cosec^{-1}x) = \frac{-1}{|x|\sqrt{1+x^2}}$ $\forall x \in \mathbb{R} - \{0\}$

Sum or difference Rule: $\frac{d}{dx}[f(x) \pm g(x)] = \frac{d}{dx}[f(x)] \pm \frac{d}{dx}[g(x)]$

Product Rule: $\frac{d}{dx}[f(x).g(x)] = f(x)\frac{d}{dx}[g(x)] + g(x)\frac{d}{dx}[f(x)]$

This is simply remembered as $(uv)' = uv' + vu'$

Quotient Rule: $\frac{d}{dx}\left[\frac{f(x)}{g(x)}\right] = \frac{g(x)\frac{d}{dx}[f(x)] - f(x)\frac{d}{dx}[g(x)]}{[g(x)]^2}$

This is simply remembered as $\left(\frac{u}{v}\right)' = \frac{vu' - uv'}{v^2}$

Chain Rule: If y is a function in t and t is a function x then $\frac{dy}{dx} = \frac{dy}{dt} \cdot \frac{dt}{dx}$

Derivative of determinant: If $y = \begin{bmatrix} f(x) & g(x) & h(x) \\ u(x) & v(x) & w(x) \\ p(x) & q(x) & r(x) \end{bmatrix}$ then

$\frac{dy}{dx} = \begin{vmatrix} f'(x) & g'(x) & h'(x) \\ u(x) & v(x) & w(x) \\ p(x) & q(x) & r(x) \end{vmatrix} + \begin{vmatrix} f(x) & g(x) & h(x) \\ u'(x) & v'(x) & w'(x) \\ p(x) & q(x) & r(x) \end{vmatrix} + \begin{vmatrix} f(x) & g(x) & h(x) \\ u(x) & v(x) & w(x) \\ p'(x) & q'(x) & r'(x) \end{vmatrix}$

Derivative of Implicit functions: Let $f(x,y) = 0$ be an implicit function. We differentiate the equation w.r.to x and take the common $\frac{dy}{dx}$ from the terms in which they have and send the rest of the terms on either side of equal and find the $\frac{dy}{dx}$

Derivative of Composite functions: Let $y = gof(x)$ then $\frac{dy}{dx} = g'(f(x)).f'(x)$

Logarithmic differentiation: Let $y = [f(x)]^{g(x)}$ then use the logarithm on both sides and we get $\log y = g(x)\log[f(x)]$. Now, we differentiate both sides using product rule and get $\frac{dy}{dx}$

Parametric differentiation: If $x = f(t)$ and $y = g(t)$ then $\frac{dy}{dx} = \frac{dy/dt}{dx/dt}$

Higher order derivatives: $\frac{d^2y}{dx^2} = \frac{d}{dx}\left(\frac{dy}{dx}\right), \frac{d^3y}{dx^3} = \frac{d}{dx}\left(\frac{d^2y}{dx^2}\right), \ldots\ldots\ldots \frac{d^ny}{dx^n} = \frac{d}{dx}\left(\frac{d^{n-1}y}{dx^{n-1}}\right)$

47. APPLICATIONS OF DERIVATIVES

Useful Mensuration Formulae:

1. Circle: Let r be the radius and d be the diameter of the circle. Then

    (i) Perimeter=$2\pi r$ **(ii)** Area=$\pi r^2 = \dfrac{\pi d^2}{4}$

2. Sector: Let r be the radius, l be the length of arc, θ be the angle made by the sector at center of the circle. Then

    (i) $l = r\theta$ **(ii)** Perimeter=$l + 2r = r(\theta + 2)$ **(iii)** Area=$\dfrac{1}{2}lr = \dfrac{1}{2}r^2\theta$

3. Square: Let a be the side of a cube. Then

    (i) Perimeter=$4a$ **(ii)** Area=a^2

4. Rectangle: Let l, b the length and breadth respectively. Then

    (i) Perimeter=$2(l + b)$ **(ii)** Area=lb

5. Cube: Let a be the side of a cube. Then

    (i) Surface area=$6a^2$ **(ii)** Volume=a^3

6. Cuboid: Let l, b, h be the length, breadth and height of the cuboid. Then

    (i) Surface area=$2(lb + bh + hl)$ **(ii)** Volume=lbh

7. Sphere: Let r be the radius of the sphere. Then

    (i) Surface area=$4\pi r^2$ **(ii)** Volume=$\dfrac{4}{3}\pi r^3$

8. Cylinder: Let r be the radius and h be the height of the cylinder. Then

    (i) Lateral surface area=$2\pi rh$ **(ii)** Total surface area=$2\pi rh + 2\pi r^2$

    (iii) Volume=$\pi r^2 h$

9. Cone: Let r be the radius, h be the height, l be the slant height and θ be the semi vertical angle. Then $l^2 = r^2 + h^2$ and $Tan\theta = \dfrac{r}{h}$

    (i) Lateral surface area=πrl **(ii)** Total surface area= $\pi rl + \pi r^2$

    (iii) Volume=$\dfrac{1}{3}\pi r^2 h$

ERRORS AND APPROXIMATIONS

Change in x: If x is a quantity, then δx is denoted as change in the x.

Change in y: If $y = f(x)$ then the change in y is denoted as δy and is defined as $\delta y = f(x + \delta x) - f(x)$

Differential of y: The differential of $y = f(x)$ is denoted by dy and is defined as $dy = f'(x)\delta x$

Approximate value: The approximate value of the function is $f(x + \delta x) = f(x) + f'(x)\delta x$

Error, relative error and percentage error:

Let $y = f(x)$ be a function with δx be the change in x. then

    (i) Absolute error in y $= \delta y$

    (ii) Relative error in y $= \frac{\delta y}{y}$

    (iii) Percentage error in y $= \frac{\delta y}{y} \times 100$

RATE MEASURE

Rate of change: If x is a variable quantity, then $\frac{dx}{dt}$ is called as rate of change of x at time t

Velocity: Let S be the displacement of a particle in time t. Then the velocity (v) is defined as the rate of change of S. i.e. $v = \frac{ds}{dt}$

    (i) The initial velocity $v = 0$

    (ii) If $v > 0$ then S increases   **(iii)** If $v < 0$ then S decreases

Acceleration: The rate of change in velocity is called as acceleration (a). i.e. $a = \frac{dv}{dt} = \frac{d^2s}{dt^2}$

Angular velocity and angular acceleration: If P is any point which moves on a curve and θ is the angle made by OP with positive X-axis in anticlockwise direction then the angular velocity of P at O is $\omega = \frac{d\theta}{dt}$ and angular acceleration $= \frac{d\omega}{dt} = \frac{d^2\theta}{dt^2}$

Velocity and acceleration in parametric form: Let $x = f(t)$ and $y = g(t)$ be a point on a curve. Then $v = \frac{ds}{dt} = \sqrt{[f'(t)]^2 + [g'(t)]^2}$ and $a = \frac{dv}{dt} = \sqrt{[f''(t)]^2 + [g''(t)]^2}$

TANGENTS AND NORMALS

Slope of a tangent and normal: If the tangent drawn to the curve $y = f(x)$ at the point $P(x_1, y_1)$ on the curve makes an angle θ with positive X-axis and anticlockwise direction then $tan\theta$ is denoted as the slope of the tangent. It is denoted by m. it is also called the gradient of the curve.

i.e. $m = tan\theta$

The straight line perpendicular to the tangent is called as the normal. So, the slope of the normal is $-\dfrac{1}{m}$

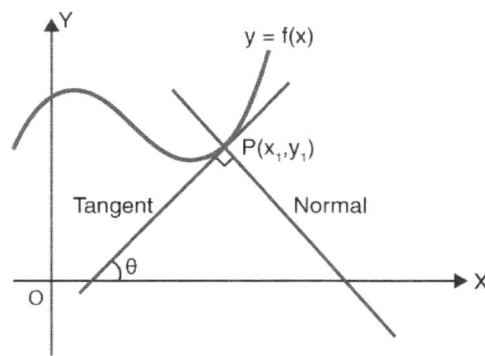

(i) The slope of the tangent of the curve $f(x, y) = 0$ at the point $P(x_1, y_1)$ is $m = -\left(\dfrac{\partial f}{\partial x} / \dfrac{\partial f}{\partial y}\right)_{(x_1, y_1)}$ and the slope of the normal is $\left(\dfrac{\partial f}{\partial y} / \dfrac{\partial f}{\partial x}\right)_{(x_1, y_1)}$

(ii) The slope of the curve $y = f(x)$ with $x = f(t)$ and $y = g(t)$ is $m = \dfrac{dy/dt}{dx/dt}$ and the slope of the normal is $-\left(\dfrac{dx/dt}{dy/dt}\right)$

(iii) The slope of the tangent parallel to the line $ax + by + c = 0$ is $m = -a/b$ and the slope of the normal is b/a

Equations of tangent and normal:

(i) The equation of the tangent to the curve $y = f(x)$ at the point (x_1, y_1) is

$$y - y_1 = m(x - x_1)$$

(ii) The equation of the normal to the curve $y = f(x)$ at the point (x_1, y_1) is

$$y - y_1 = -\dfrac{1}{m}(x - x_1)$$

Length of tangent, normal, sub-tangent, sub-normal: Let the tangent and normal drawn to the curve $y = f(x)$ at the point $P(x_1, y_1)$ meets the X-axis at T and N respectively. Let PM be the perpendicular to X-axis. Let m be the slope of the tangent. Then

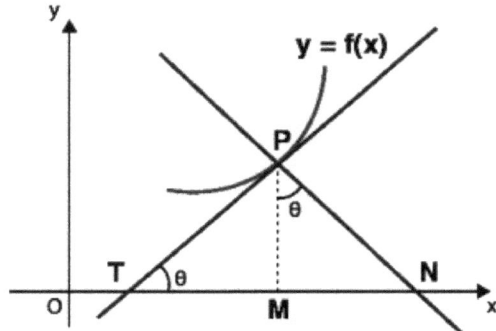

(i) Length of the tangent = PT = $\left|\dfrac{y_1\sqrt{1+m^2}}{m}\right|$

(ii) Length of the normal = PN = $\left|y_1\sqrt{1+m^2}\right|$

(iii) Length of the sub-tangent = TM = $\left|\dfrac{y_1}{m}\right|$

(iv) Length of the sub-normal = MN = $|y_1 m|$

Angle between two curves: The angle between the two curves is the angle between the tangents drawn at the intersecting points of the curves.

Let $P(x_1, y_1)$ be the point of intersection of the curves $y = f(x)$ and $y = g(x)$. Let m_1 and m_2 be the slopes of the tangents drawn to the curves at P. Then the angle between the curves θ is given by $Tan\theta = \left|\dfrac{m_1 - m_2}{1 + m_1 m_2}\right|$

(i) Two curves touch each other when $m_1 = m_2$

(ii) Two curves intersects orthogonally when $m_1 m_2 = -1$

MEAN VALUE THEOREMS

Rolle's Theorem: Let the function $f: [a, b] \to \mathcal{R}$ is such that

  (i) f is continuous on $[a, b]$

  (ii) f is differentiable on (a, b)

  (iii) $f(a) = f(b)$

Then there exists at least one value c of x in the interval (a, b) such that $f'(c) = 0$

Legrange's Mean Value Theorem: Let the function $f: [a, b] \to \mathcal{R}$ is such that

  (i) f is continuous on $[a, b]$

  (ii) f is differentiable on (a, b)

Then there exists at least one value c of x in the interval (a, b) such that

$$f'(c) = \dfrac{f(b) - f(a)}{b - a}$$

Intermediate value theorem: Let $f(x)$ be continuous on an interval $[a, b]$. If y_0 is a number between $f(a)$ and $f(b)$ then there exists a number c between a and b such that

$$f(c) = y_0 = \frac{f(a) + f(b)}{2}$$

Cauchy's Mean Value Theorem: Let the functions $f : [a, b] \to \mathcal{R}$ and $\emptyset : [a, b] \to \mathcal{R}$ is such that

    (i) both are continuous on $[a, b]$

    (ii) both are differentiable on (a, b)

    (iii) $\emptyset'(x) \neq 0$ for any value of x in the interval (a, b)

Then there exists at least one value c of x in the interval (a, b) such that

$$\frac{f'(c)}{\emptyset'(c)} = \frac{f(b) - f(a)}{\emptyset(b) - \emptyset(a)}$$

MAXIMA AND MINIMA

Increasing function: Let f be a real valued function having domain D with $(a, b) \subseteq D$. Then f is said to be an increasing function on (a, b) when

$$x_1 < x_2 \Rightarrow f(x_1) \leq f(x_2) \; \forall x_1, x_2 \in (a, b)$$

Decreasing function: Let f be a real valued function having domain D with $(a, b) \subseteq D$. Then f is said to be a decreasing function on (a, b) when

$$x_1 < x_2 \Rightarrow f(x_1) \geq f(x_2) \; \forall x_1, x_2 \in (a, b)$$

Strictly Increasing function: Let f be a real valued function having domain D with

    $(a, b) \subseteq D$. Then f is said to be a strictly increasing function on (a, b) when

$$x_1 < x_2 \Rightarrow f(x_1) < f(x_2) \; \forall x_1, x_2 \in (a, b)$$

Strictly Decreasing function: Let f be a real valued function having domain D with

    $(a, b) \subseteq D$. Then f is said to be a strictly decreasing function on (a, b) when

$$x_1 < x_2 \Rightarrow f(x_1) > f(x_2) \; \forall x_1, x_2 \in (a, b)$$

Monotonic Function: A function which is either increasing or decreasing on its domain is called as a monotonic function.

Test for Monotonicity:

    (i) f is increasing on (a, b) when $f'(x) \geq 0 \; \forall x \in (a, b)$

    (ii) f is decreasing on (a, b) when $f'(x) \leq 0 \; \forall x \in (a, b)$

(iii) f is strictly increasing on (a,b) when $f'(x) > 0$ $\forall x \in (a,b)$

(iv) f is strictly decreasing on (a,b) when $f'(x) < 0$ $\forall x \in (a,b)$

Critical point: Let f be a real valued function having domain D. Then a point $x = a \in D$ is said to be critical point when $f'(a) = 0$ or $f'(a)$ does not exist.

Stationary point: If $f'(a) = 0$ then $f(a)$ is called stationary value of f at $x = a$ and $(a, f(a))$ is called as a stationary point of f.

Maxima or Absolute maxima or global maxima or greatest value:

Let $f(x)$ be a function with domain D then $f(x)$ is said to be have maximum value at a point $a \in D$ when $f(x) \leq f(a)$ $\forall x \in D$.

a is called the point of maxima and $f(a)$ is called maximum value of $f(x)$

Minima or Absolute minima or global minima or least value:

Let $f(x)$ be a function with domain D then $f(x)$ is said to be have minimum value at a point $a \in D$ when $f(x) \geq f(a)$ $\forall x \in D$.

a is called the point of minima and $f(a)$ is called minimum value of $f(x)$

Local maxima and Local minima:

Let $f(x)$ be a function. Then $f(x)$ is said to be have a local maximum at a point $x = a$ when $f(x) \leq f(a)$ $\forall x \in (a - h, a + h)$ where h is a small positive quantity. $f(a)$ is called local maximum.

Let $f(x)$ be a function. Then $f(x)$ is said to be have a local minimum at a point $x = a$ when $f(x) \geq f(a)$ $\forall x \in (a - h, a + h)$ where h is a small positive quantity. $f(a)$ is called local minimum.

Extreme points and Extreme values: If a function $f(x)$ has a local maximum and local minimum at the points $x = a$ and $x = b$ respectively then the points are called as extreme points and $f(a), f(b)$ are called as extreme values.

First derivative test: Let $f(x)$ be a differential function on an interval I and $a \in I$. Then

$f(x)$ has local maximum or minimum at $x = a$ when $f'(a) = 0$

Second derivative test: Let $f(x)$ be a differential function on an interval I and $a \in I$. Then

    (i) $f(x)$ has local maximum at $x = a$ when $f'(a) = 0$ and $f''(a) < 0$

    (ii) $f(x)$ has local minimum at $x = a$ when $f'(a) = 0$ and $f''(a) > 0$

Global maximum and Global minimum in $[a, b]$: Let $c_1, c_2, \ldots \ldots c_n$ be the critical points of the function $f(x)$ in (a, b). Then

    (i) Global maximum=$Maximum\ of\ \{f(a), f(c_1), f(c_2), \ldots.. f(c_n), f(b)\}$

    (ii) global minimum= $Minimum\ of\ \{f(a), f(c_1), f(c_2), \ldots.. f(c_n), f(b)\}$

Global maximum and Global minimum in (a, b): Let $c_1, c_2, \ldots\ldots c_n$ be the critical points of the function $f(x)$ in (a, b). Then

(i) Global maximum = $Maximum\ of\ \{f(c_1), f(c_2), \ldots..f(c_n)\}$

(ii) global minimum = $Minimum\ of\ \{f(c_1), f(c_2), \ldots..f(c_n)\}$

Convex function theorem: Let $f''(x) < 0$ in $[a, b]$. Then

(i) $\frac{f(a)+f(b)}{2} \leq f\left(\frac{a+b}{2}\right)$ (ii) $\frac{\alpha f(a)+\beta f(b)}{\alpha+\beta} \leq f\left(\frac{\alpha a+\beta b}{\alpha+\beta}\right)$ where $\alpha > 0, \beta > 0$

(iii) $\frac{f(a_1)+f(a_2)+\cdots..+f(a_n)}{n} \leq f\left(\frac{a_1+a_2+\cdots..+a_n}{n}\right)$

Concave function theorem: Let $f''(x) > 0$ in $[a, b]$. Then

(i) $\frac{f(a)+f(b)}{2} \geq f\left(\frac{a+b}{2}\right)$ (ii) $\frac{\alpha f(a)+\beta f(b)}{\alpha+\beta} \geq f\left(\frac{\alpha a+\beta b}{\alpha+\beta}\right)$ where $\alpha > 0, \beta > 0$

(iii) $\frac{f(a_1)+f(a_2)+\cdots..+f(a_n)}{n} \geq f\left(\frac{a_1+a_2+\cdots..+a_n}{n}\right)$

48. INDEFINITE INTEGRATION

Integration: Let $F(x)$ be a differential function of x then the integration of $f(x)$ w.r.to x is defined as $\int f(x)dx = F(x) + c$ where c is an integral constant.

The integration is also called as primitive or anti derivative.

Properties of integral:

(i) $\int [f(x) \pm g(x)]dx = \int f(x)dx \pm \int g(x)dx + c$

(ii) $\int [f_1(x) \pm f_2(x) \pm \ldots \pm f_n(x)]dx = \int f_1(x)dx \pm \int f_2(x)dx \pm \ldots \int f_n(x)dx + c$

(iii) $\int kf(x)dx = k\int f(x)dx + c$

Important Formulae:

(i) $\int \frac{d}{dx}[f(x)]dx = f(x) + c$

(ii) $\int dx = x + c$

(iii) $\int k\,dx = kx + c$

(iv) $\int x^n dx = \frac{x^{n+1}}{n+1} + c$

(v) $\int \frac{1}{\sqrt{x}}dx = 2\sqrt{x} + c$

(vi) $\int \sqrt{x}\,dx = \frac{2}{3}x^{3/2} + c$

(vii) $\int \frac{1}{x^2}dx = -\frac{1}{x} + c$

(viii) $\int \frac{1}{x}dx = \log|x| + c$

(ix) $\int e^x dx = e^x + c$

(x) $\int a^x dx = \frac{a^x}{\log a} + c$

(xi) $\int Sinx\, dx = -Cosx + c$

(xii) $\int Cosx\, dx = Sinx + c$

(xiii) $\int Tanx\, dx = \log|Secx| + c$

(xiv) $\int Cotx\, dx = \log|Sinx| + c$

(xv) $\int Secx\, dx = \log|Secx + Tanx| + c$ (or) $\log\left|Tan\left(\frac{\pi}{4} + \frac{x}{2}\right)\right| + c$

(xvi) $\int Cosecx\, dx = \log|Cosecx - Cotx| + c$ (or) $\log\left|Tan\frac{x}{2}\right| + c$

(xvii) $\int Sec^2x\, dx = Tanx + c$

(xviii) $\int Cosec^2x\, dx = -Cotx + c$

MATHEMATICS SUCCESS MANTRA

(xix) $\int Secx.Tanx\, dx = Secx + c$

(xx) $\int Cosecx.Cotx\, dx = -Cosecx + c$

(xxi) $\int Sinhx\, dx = Coshx + c$

(xxii) $\int Coshx\, dx = Sinhx + c$

(xxiii) $\int Tanhx\, dx = log|Coshx| + c$

(xxiv) $\int Cothx\, dx = log|Sinhx| + c$

(xxv) $\int Sech^2 x\, dx = Tanhx + c$

(xxvi) $\int Cosech^2 x\, dx = -Cothx + c$

(xxvii) $\int Sechx.Tanhx\, dx = -Sechx + c$

(xxviii) $\int Cosechx.Cothx\, dx = -Cosechx + c$

(xxix) $\int \frac{dx}{\sqrt{1-x^2}} = Sin^{-1}x + c = -Cos^{-1}x + c$

(xxx) $\int \frac{dx}{1+x^2} = Tan^{-1}x + c = -Cot^{-1}x + c$

(xxxi) $\int \frac{dx}{|x|\sqrt{x^2-1}} = Sec^{-1}x + c = -Cosec^{-1}x + c$

(xxxii) $\int \frac{dx}{\sqrt{1+x^2}} = Sinh^{-1}x + c = log(x + \sqrt{x^2+1}) + c$

(xxxiii) $\int \frac{dx}{\sqrt{x^2-1}} = Cosh^{-1}x + c = log(x + \sqrt{x^2-1}) + c$

(xxxiv) $\int \frac{dx}{\sqrt{a^2-x^2}} = Sin^{-1}\left(\frac{x}{a}\right) + c$

(xxxv) $\int \frac{dx}{\sqrt{x^2-a^2}} = Cosh^{-1}\left(\frac{x}{a}\right) + c = log(x + \sqrt{x^2-a^2}) + c$

(xxxvi) $\int \frac{dx}{\sqrt{x^2+a^2}} = Sinh^{-1}\left(\frac{x}{a}\right) + c = log(x + \sqrt{x^2+a^2}) + c$

(xxxvii) $\int \frac{dx}{a^2+x^2} = \frac{1}{a}Tan^{-1}\left(\frac{x}{a}\right) + c$

(xxxviii) $\int \frac{dx}{a^2-x^2} = \frac{1}{2a}log\left|\frac{a+x}{a-x}\right| + c$

(xxxix) $\int \frac{dx}{x^2-a^2} = \frac{1}{2a}log\left|\frac{x-a}{x+a}\right| + c$

(xl) $\int \sqrt{a^2-x^2}\, dx = \frac{x}{2}\sqrt{a^2-x^2} + \frac{a^2}{2}Sin^{-1}\left(\frac{x}{a}\right) + c$

(xli) $\int \sqrt{a^2+x^2}\, dx = \frac{x}{2}\sqrt{a^2+x^2} + \frac{a^2}{2}Sinh^{-1}\left(\frac{x}{a}\right) + c$

(xlii) $\int \sqrt{x^2-a^2}\, dx = \frac{x}{2}\sqrt{x^2-a^2} - \frac{a^2}{2}Cosh^{-1}\left(\frac{x}{a}\right) + c$

(xliii) $\int [f(x)]^n f'(x)\, dx = \frac{[f(x)]^{n+1}}{n+1} + c$

(xliv) $\int \frac{f'(x)}{\sqrt{f(x)}}\, dx = 2\sqrt{f(x)} + c$

(xlv) $\int \frac{f'(x)}{f(x)} dx = \log|f(x)| + c$

(xlvi) $\int \sqrt{f(x)}\, f'(x) dx = \frac{2}{3}[f(x)]^{3/2} + c$

(xlvii) $\int e^x[f(x) + f'(x)] dx = e^x . f(x) + c$

Integral by parts: Let $f(x)$ and $g(x)$ be two functions. Then

$$\int f(x).g(x)dx = f(x) \int g(x)dx - \int \left[f'(x) \int g(x)dx\right] dx$$

This simply remembered as $\int u\, dv = uv - \int v\, du$

$f(x)$ should be taken by the priority of letters from the word **ILATE**

    I stands for inverse trigonometric function

    L stands for logarithmic function

    A stands for algebraic function

    T stands for trigonometric function

    E stands for exponential function

Methods to evaluate integrations:

(i) $\int \sqrt{ax^2 + bx + c}\, dx$ or $\int \frac{1}{\sqrt{ax^2+bx+c}}\, dx$ or $\int \frac{1}{ax^2+bx+c}\, dx$

Convert $ax^2 + bx + c$ into the form of sum or difference of perfect squares of two terms and then integrate to evaluate the integration

(ii) $\int \frac{px+q}{ax^2+bx+c}\, dx$ or $\int (px+q)\sqrt{ax^2 + bx + c}\, dx$ or $\int \frac{px+q}{\sqrt{ax^2+bx+c}}\, dx$

Let $px + q = A\frac{d}{dx}(ax^2 + bx + c) + B$ and solve for the values A and B and then integrate to evaluate the integration

(iii) $\int \frac{1}{(px+q)\sqrt{ax^2+bx+c}}\, dx$

Let $px + q = \frac{1}{t}$ and substitute the value of x in given integral and integrate to evaluate the integration

(iv) $\int (ax + b)\sqrt{px + q}\, dx$ or $\int \frac{\sqrt{px+q}}{ax+b}\, dx$ or $\int \frac{1}{(ax+b)\sqrt{px+q}}\, dx$ or $\int \frac{1}{(ax^2+bx+c)\sqrt{px+q}}\, dx$

Let $px + q = t^2$ and substitute the value of x in given integral and integrate to evaluate the integration

(v) $\int \frac{1}{a+b\sin x}\, dx$ or $\int \frac{1}{a+b\cos x}\, dx$ or $\int \frac{1}{a\cos x+b\sin x+c}\, dx$

Let $\tan \frac{x}{2} = t$ then $\sin x = \frac{2t}{1+t^2}$; $\cos x = \frac{1-t^2}{1+t^2}$; $dx = \frac{2dt}{1+t^2}$

substitute these values in given integral and integrate to evaluate the integration

(vi) $\int \frac{1}{a+b\sin 2x}\, dx$ or $\int \frac{1}{a+b\cos 2x}\, dx$ or $\int \frac{1}{a\cos 2x+b\sin 2x+c}\, dx$

Let $\tan x = t$ then $\sin 2x = \frac{2t}{1+t^2}$; $\cos 2x = \frac{1-t^2}{1+t^2}$; $dx = \frac{dt}{1+t^2}$

substitute these values in given integral and integrate to evaluate the integration

(vii) $\int \frac{a\cos x+b\sin x}{c\cos x+d\sin x}\, dx$

Let $a\cos x + b\sin x = A\frac{d}{dx}(c\cos x + d\sin x) + B(c\cos x + d\sin x)$

solve for the values A and B and then integrate to evaluate the integration

(viii) $\int \frac{a\cos x+b\sin x+c}{d\cos x+e\sin x+f}\, dx$

Let $a\cos x + b\sin x = A\frac{d}{dx}(c\cos x + d\sin x) + B(c\cos x + d\sin x) + C$

solve for the values A, B and C and then integrate to evaluate the integration

(ix) $\int \frac{1}{a+b\sin^2 x}\, dx$ or $\int \frac{1}{a+b\cos^2 x}\, dx$ or $\int \frac{1}{a\cos^2 x+b\sin^2 x+c}\, dx$ or $\int \frac{1}{a\cos^2 x+b\sin^2 x+c\cos x\sin x}\, dx$ or $\int \frac{1}{(a\cos x+b\sin x)^2}\, dx$

Divide numerator and denominator with $\cos^2 x$ and then take $\tan x = t$.

substitute these values in given integral and integrate to evaluate the integration

Reduction Formulae:

(i) If $I_n = \int x^n e^{ax} dx$ then $I_n = \dfrac{e^{ax}}{a} x^n - \dfrac{n}{a} I_{n-1}$ for $n \geq 1$

(ii) If $I_n = \int Sin^n x \, dx$ then $I_n = \dfrac{-Sin^{n-1}x.Cosx}{n} + \dfrac{n-1}{n} I_{n-2}$ for $n \geq 2$

(iii) If $I_n = \int Cos^n x \, dx$ then $I_n = \dfrac{Cos^{n-1}x.Sinx}{n} + \dfrac{n-1}{n} I_{n-2}$ for $n \geq 2$

(iv) If $I_n = \int Tan^n x \, dx$ then $I_n = \dfrac{Tan^{n-1}x}{n-1} - I_{n-2}$ for $n \geq 2$

(v) If $I_n = \int Cot^n x \, dx$ then $I_n = \dfrac{-Cot^{n-1}x}{n-1} - I_{n-2}$ for $n \geq 2$

(vi) If $I_n = \int Sec^n x \, dx$ then $I_n = \dfrac{Sec^{n-2}x.Tanx}{n-1} + \dfrac{n-2}{n-1} I_{n-2}$ for $n \geq 2$

(vii) If $I_n = \int Cosec^n x \, dx$ then $I_n = \dfrac{-Cosec^{n-2}x.Cotx}{n-1} + \dfrac{n-2}{n-1} I_{n-2}$ for $n \geq 2$

(vi) If $I_{m,n} = \int Sin^m x . Cos^n x \, dx$ then $I_n = \dfrac{-Sin^{m-1}x.Cos^{n+1}x}{m+n} + \dfrac{m-1}{m+n} I_{m-2,n}$ for $m \geq 2$

(or) $\dfrac{Sin^{m+1}x.Cos^{n-1}x}{m+n} + \dfrac{n-1}{m+n} I_{m,n-2}$ for $n \geq 2$

49. DEFINITE INTEGRALS

Definite Integral: Let $f(x)$ be a function defined on $[a, b]$. If $F(x)$ is the integral of $f(x)$ w.r.to x then the definite integral of $f(x)$ is defined as

$$\int_a^b f(x)dx = F(b) - F(a).$$

a is called the lower limit and b is called the upper limit.

Properties:

(i) $\int_a^b f(x)dx = -\int_b^a f(x)dx$

(ii) $\int_a^b f(x)dx = \int_a^c f(x)dx + \int_c^b f(x)dx$ for $a < c < b$

(iii) $\int_0^a f(x)dx = \int_0^a f(a-x)dx$

(iv) $\int_a^b f(x)dx = \int_a^b f(a+b-x)dx$

(v) $\int_{-a}^a f(x)dx = \begin{cases} 2\int_0^a f(x)dx & \text{when } f \text{ is even} \\ 0 & \text{when } f \text{ is odd} \end{cases}$

(vi) $\int_0^{2a} f(x)dx = \begin{cases} 2\int_0^a f(x)dx & \text{when } f(2a-x) = f(x) \\ 0 & \text{when } f(2a-x) = -f(x) \end{cases}$

(vii) If $f(x) \geq 0$ in $[a, b]$ then $\int_a^b f(x)dx \geq 0$

(viii) If $f(x) \leq g(x)$ in $[a, b]$ then $\int_a^b f(x)dx \leq \int_a^b g(x)dx$

(ix) If $f(x)$ is a periodic function with period T then

$$\int_0^{nT} f(x)dx = n\int_0^T f(x)dx$$

$$\int_0^{a+nT} f(x)dx = n\int_0^{a+T} f(x)dx$$

$$\int_{a+nT}^{b+nT} f(x)dx = \int_a^b f(x)dx$$

Leibnitz Rule:

$$\frac{d}{dx}\left[\int_{\varphi(x)}^{\mu(x)} f(t)dt\right] = f[\mu(x)]\frac{d}{dx}[\mu(x)] - f[\varphi(x)]\frac{d}{dx}[\varphi(x)]$$

Mean value of a function: Let $f(x)$ be a continuous function defined on $[a, b]$. Then there exists a point $c \in (a, b)$ such that $\int_a^b f(x)dx = f(c)(b-a)$ then $f(c) = \frac{1}{b-a}\int_a^b f(x)dx$ is called as the mean value of $f(x)$ over $[a, b]$

Reduction Formulae:

(i) $I_n = \int_0^{\pi/2} Sin^n x \, dx = \int_0^{\pi/2} Cos^n x \, dx = \begin{cases} \frac{n-1}{n} \cdot \frac{n-3}{n-2} \cdot \frac{n-5}{n-4} \cdots \frac{1}{2} \cdot \frac{\pi}{2} & \text{when } n \text{ is even} \\ \frac{n-1}{n} \cdot \frac{n-3}{n-2} \cdot \frac{n-5}{n-4} \cdots \frac{2}{3} & \text{when } n \text{ is odd} \end{cases}$

(ii) $I_n = \int_0^{\pi/4} Tan^n x \, dx = \begin{cases} \frac{1}{n-1} - \frac{1}{n-3} + \frac{1}{n-5} - \cdots \cdots \frac{\pi}{4} & \text{when } n \text{ is even} \\ \frac{1}{n-1} - \frac{1}{n-3} + \frac{1}{n-5} - \cdots \cdots \frac{1}{2} log 2 & \text{when } n \text{ is odd} \end{cases}$

(iii) $I_{m,n} = \int_0^{\pi/2} Sin^m x \cdot Cos^n x \, dx =$
$\begin{cases} \frac{n-1}{m+n} \cdot \frac{n-3}{m+n-2} \cdots \cdots \frac{2}{m+3} \cdot \frac{1}{m+1} & \text{when } n \neq 1 \text{ is odd} \\ \frac{n-1}{m+n} \cdot \frac{n-3}{m+n-2} \cdots \frac{1}{m+2} \cdot \frac{m-1}{m} \cdot \frac{m-3}{m-2} \cdots \cdots \frac{1}{2} \cdot \frac{\pi}{2} & \text{when } n \text{ is even and } m \text{ is even} \\ \frac{n-1}{m+n} \cdot \frac{n-3}{m+n-2} \cdots \frac{1}{m+2} \cdot \frac{m-1}{m} \cdot \frac{m-3}{m-2} \cdots \cdots \frac{2}{3} & \text{when } n \text{ is even and } m \neq 1 \text{ is odd} \end{cases}$

Definite integral as a limit of a sum:

If $f(x)$ is a continuous function on $[0,1]$ and $P = \left\{0, \frac{1}{n}, \frac{2}{n}, \frac{3}{n}, \ldots \frac{n-1}{n}, 1\right\}$ is a partition of $[0,1]$ into n sub-intervals each of length, $\frac{1}{n}$ then the definite integral is given by

$$\int_0^1 f(x) dx = \lim_{n \to \infty} \frac{1}{n} \sum_{i=1}^{n} f\left(\frac{i}{n}\right)$$

Extension this formula for the interval $[0, p]$, it is given by

$$\int_0^p f(x) dx = \lim_{n \to \infty} \frac{1}{n} \sum_{i=1}^{np} f\left(\frac{i}{n}\right)$$

50. AREAS

1. The area bounded by the continuous curve $y = f(x) \geq 0$, X-axis and $x = a$; $x = b$ $(a < b)$ is given by $A = \int_a^b f(x)dx$

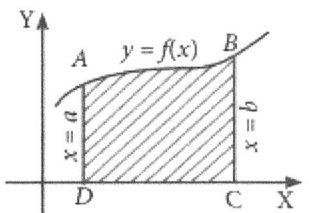

2. The area bounded by the continuous curve $y = f(x) \leq 0$, X-axis and $x = a$; $x = b$ $(a < b)$ is given by $A = -\int_a^b f(x)dx$

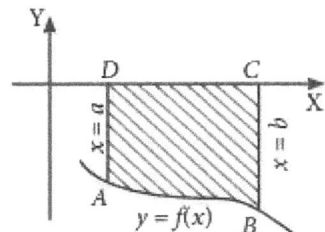

3. The area bounded by the continuous curve $x = f(y) \geq 0$, Y-axis and $y = c$; $y = d$ $(c < d)$ is given by $A = \int_c^d f(y)dy$

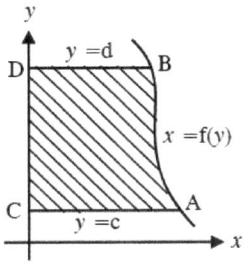

4. The area bounded by the continuous curve $x = f(y) \leq 0$, Y-axis and $y = c$; $y = d$ $(c < d)$ is given by $A = -\int_c^d f(y)dy$

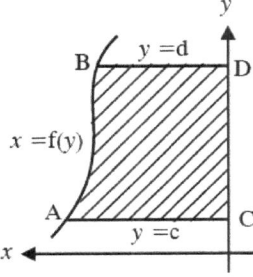

5. If the curve $y = f(x)$ cross the X-axis at $(c, 0)$ then the area bounded by the curve and $x = a$; $x = b$ with $(a < b ; a < c < b)$ is given by $A = \int_a^c f(x)dx - \int_c^b f(x)dx$

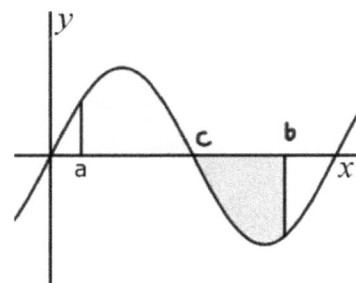

6. If two curves $y = f(x)$ and $y = g(x)$ such that $f(x) \leq g(x)$ intersect at two points $x = a$; $x = b$ $(a < b)$ then the area bounded by the curves is given by $A = \int_a^b [g(x) - f(x)]dx$

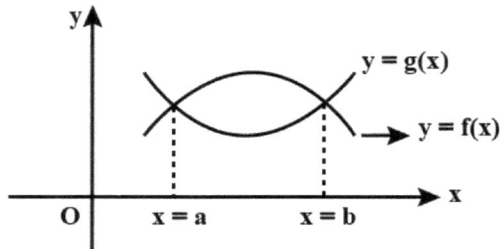

7. If two curves $y = f(x)$ and $y = g(x)$ are intersect at a point $x = c$ such that $f(x) \geq g(x)$ for $x \in [a, c]$ and $f(x) \leq g(x)$ for $x \in [c, b]$ then the area bounded by the curves is given by $A = \int_a^c [f(x) - g(x)]dx + \int_c^b [g(x) - f(x)]dx$

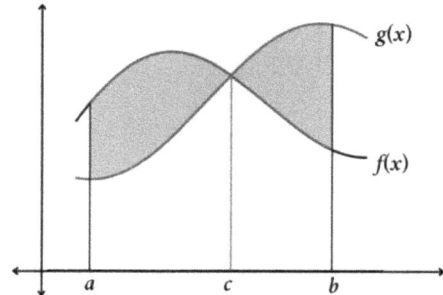

51. DIFFERENTIAL EQUATIONS

Differential Equation: The equation involving one dependent variable and its derivatives w.r.to one or more independent variables is called as a differential equation.

Eg: $\frac{d^3x}{dx^3} + 5\frac{d^2x}{dx^2} - 3\frac{dy}{dx} + 4y = 2x^2$

Types of differential equations:

(i) **Ordinary differential equation:** The differential equation involving derivatives wr.to single independent variable is called as an ordinary differential equation.

Eg: $\frac{dy}{dx} + 4x = 2x^2$

(ii) **Partial differential equation:** The differential equation involving at least two independent variables and partial derivatives wr.to either of these independent variables is called as a partial differential equation.

Eg: $y\frac{\partial u}{\partial x} + 4x\frac{\partial u}{\partial y} = 7u$

Order of a differential equation: The order of a differential equation is defined as the order of highest order derivative in the equation.

Eg: The order of $\left(\frac{d^3x}{dx^3}\right)^2 + 5\left(\frac{d^2x}{dx^2}\right)^4 - 3\frac{dy}{dx} + 4y = 2x^2$ is 3

Degree of a differential equation: The degree of a differential equation is defined as the index of highest order derivative in the equation.

Eg: The degree of $\left(\frac{d^3x}{dx^3}\right)^2 + 5\left(\frac{d^2x}{dx^2}\right)^4 - 3\frac{dy}{dx} + 4y = 2x^2$ is 2

Formation of a differential equation: By eliminating the arbitrary constants in the given general equation through differentiation is said to be formation of differential equation.

The order of the differential equation is the number of arbitrary constants in the original general equation.

Solutions of differential equations:

(i) **General solution:** The solution which contains as many arbitrary constants as the order of the differential equation is called as a general solution of the differential equation.

(ii) **Particular solution:** The solution which contains the particular values for the arbitrary constants as the order of the differential equation is called as a general solution of the differential equation.

Methods to solve the differential equations:

(i) Variable-Separable method: In this method, separate the variables and its derivatives as one side and integrating to get the solution.

(ii) Method to solve homogeneous equations: The equation is in the form $\frac{dy}{dx} = \frac{f(x,y)}{g(x,y)}$ such that $\frac{f(kx,ky)}{g(kx,ky)} = k^\alpha \frac{f(x,y)}{g(x,y)}$ for some α is called as a homogeneous equation.

To solve the homogeneous equations, take $y = vx$ and $\frac{dy}{dx} = v + x\frac{dv}{dx}$. Substitute these values in the given equation and then solve the equation through variable-separable method.

(iii) Method to solve nonhomogeneous equations:

(a) The equation is in the form $\frac{dy}{dx} = \frac{a_1x+b_1y+c_1}{a_2x+b_2y+c_2}$ where $b_1 = -a_2$ is called as the nonhomogeneous equation of first type.

To solve these equations, cross multiplying the equation and get the terms to integrate. Directly integrating and get the solution easily.

(b) The equation is in the form $\frac{dy}{dx} = \frac{a_1x+b_1y+c_1}{a_2x+b_2y+c_2}$ where $\frac{a_1}{a_2} = \frac{b_1}{b_2}$ and $b_1 \neq -a_2$ is called as the nonhomogeneous equation of second type.

To solve these equations, let $a_1x + b_1y = t$ and convert the equation in terms of t and x. solve the equation through variable-separable method.

(c) The equation is in the form $\frac{dy}{dx} = \frac{a_1x+b_1y+c_1}{a_2x+b_2y+c_2}$ where $\frac{a_1}{a_2} \neq \frac{b_1}{b_2}$ and $b_1 \neq -a_2$ is called as the nonhomogeneous equation of third type.

To solve these equations, let $x = X + h$ and $y = Y + k$. Substitute these values in the equation and convert the equation in the form of homogeneous equation $\frac{dY}{dX} = \frac{a_1X+b_1Y}{a_2X+b_2Y}$ by taking $a_1h + b_1k + c_1 = 0$ and $a_2h + b_2k + c_2 = 0$ and solve for the values h and k. Solve the equation by method to solve homogeneous equations.

(iv) Method to solve linear equations: The equation is in the form of $\frac{dy}{dx} + Py = Q$ where P and Q are the functions in the variable x.

To solve these equations, find integral factor using the formula $I.F. = e^{\int P\,dx}$ and then get the solution through the equation $y(I.F.) = \int Q\,(I.F.)\,dx$

Another form of linear equation is of $\frac{dx}{dy} + Px = Q$ where P and Q are the functions in the variable y.

To solve these equations, find integral factor using the formula $I.F. = e^{\int P\, dy}$ and then get the solution through the equation $x(I.F.) = \int Q\,(I.F.)\, dy$

(iv) Method to solve Bernoulli's equations:

The equation is in the form of $\frac{dy}{dx} + Py = Qy^n$ where P and Q are the functions in the variable x.

To solve these equations, divide the equation with y^n and take $\frac{1}{y^{n-1}} = z$. Then convert the equation into linear equation and solve the equation by method to solve the linear equation.

Another form of Bernoulli's equation is of $\frac{dx}{dy} + Px = Qx^n$ where P and Q are the functions in the variable y.

To solve these equations, divide the equation with x^n and take $\frac{1}{x^{n-1}} = z$. Then convert the equation into linear equation and solve the equation by method to solve the linear equation.

Orthogonal Trajectory: Any curve cuts every member of curve at right angles is called as an orthogonal trajectory of the family.

To solve these equations, convert the equation into the form $f(x, y, c) = 0$. Differentiate the equation and eliminate the parameter c. Now, replace $\frac{dy}{dx}$ with $-\frac{dx}{dy}$ and solve the differential equation. The solution is the orthogonal trajectory of the original equation.

STATISTICS

52. MEASURES OF DISPERSION

Arithmetic mean:

(i) Individual series: Let $x_1, x_2, x_3, \ldots \ldots, x_n$ be the values of the variable x. Then

$$\bar{x} = \frac{x_1 + x_2 + x_3 + \cdots \ldots + x_n}{n} = \frac{1}{n}\sum_{i=1}^{n} x_i$$

(or) $\bar{x} = A + \frac{1}{n}\sum_{i=1}^{n}(x_i - A)$ where A is the assumed mean.

(ii) Discrete series: Let $x_1, x_2, x_3, \ldots \ldots, x_n$ be the values of the variable x and $f_1, f_2, f_3, \ldots \ldots, f_n$ be the corresponding frequencies then

$$\bar{x} = \frac{f_1 x_1 + f_2 x_2 + f_3 x_3 + \cdots \ldots + f_n x_n}{N} = \frac{1}{N}\sum_{i=1}^{n} f_i x_i \text{ where } N = \sum_{i=1}^{n} f_i$$

(iii) Continuous series: Let $x_1, x_2, x_3, \ldots \ldots, x_n$ be the mid values of the class intervals and $f_1, f_2, f_3, \ldots \ldots, f_n$ be the corresponding frequencies then

$$\bar{x} = \frac{f_1 x_1 + f_2 x_2 + f_3 x_3 + \cdots \ldots + f_n x_n}{N} = \frac{1}{N}\sum_{i=1}^{n} f_i x_i \text{ where } N = \sum_{i=1}^{n} f_i$$

Weighted Arithmetic mean: Let $w_1, w_2, w_3, \ldots \ldots, w_n$ be the weights assigned to the values $x_1, x_2, x_3, \ldots \ldots, x_n$ respectively of a variable x then the weighted mean is

$$\bar{x} = \frac{w_1 x_1 + w_2 x_2 + w_3 x_3 + \cdots \ldots + w_n x_n}{N} = \frac{1}{N}\sum_{i=1}^{n} w_i x_i \text{ where } N = \sum_{i=1}^{n} w_i$$

Geometric mean:

(i) Individual series: Let $x_1, x_2, x_3, \ldots \ldots, x_n$ be the values of the variable x. Then
$G.M. = (x_1 . x_2 . x_3 . \ldots \ldots x_n)^{1/n}$

(ii) Discrete series (or) Continuous series: Let $x_1, x_2, x_3, \ldots \ldots, x_n$ be the values of the variable x and $f_1, f_2, f_3, \ldots \ldots, f_n$ be the corresponding frequencies then

$$G.M. = (x_1^{f_1} . x_2^{f_2} . x_3^{f_3} . \ldots \ldots x_n^{f_n})^{1/N} \text{ where } N = \sum_{i=1}^{n} f_i$$

Harmonic mean:

(i) Individual series: Let $x_1, x_2, x_3, \ldots, x_n$ be the values of the variable x. Then

$$H.M. = \frac{1}{\frac{1}{n}\left(\frac{1}{x_1} + \frac{1}{x_2} + \cdots + \frac{1}{x_n}\right)} \quad (or) \quad \frac{1}{H.M.} = \frac{1}{n}\sum_{i=1}^{n}\frac{1}{x_i}$$

(ii) Discrete series (or) Continuous series: Let $x_1, x_2, x_3, \ldots, x_n$ be the values of the variable x and $f_1, f_2, f_3, \ldots, f_n$ be the corresponding frequencies then

$$\frac{1}{H.M.} = \frac{1}{n}\sum_{i=1}^{n} f_i \frac{1}{x_i} \text{ where } N = \sum_{i=1}^{n} f_i$$

Median:

(i) Individual series: If the items are arranged in ascending or descending order of magnitude then the median is the $\frac{n+1}{2}^{th}$ item, in the case of odd number of items and average of $\frac{n}{2}^{th}$ and $\frac{n+2}{2}^{th}$ items, in the case of even number of items.

(ii) Discrete series: Arrange the items in ascending or descending order and find the cumulative frequencies of the items. Then

$$\text{Median} = \begin{cases} \frac{N+1}{2}^{th} \text{ item when N is odd} \\ \text{Average of } \frac{n}{2}^{th} \text{ and } \frac{n+2}{2}^{th} \text{ items when N is even} \end{cases}$$

where $N = \sum_{i=1}^{n} f_i$

(iii) Continuous series:

$$Median = l + \frac{\left(\frac{N}{2} - F\right)C.I.}{f}$$

Where $\frac{N}{2}$ = the median class where $N = \sum_{i=1}^{n} f_i$

l = the lower limit of the median class

f = the frequency of the median class

F = the cumulative frequency of the class preceding the median class

$C.I.$ = the class interval of the median class

Mode:

(i) Individual series: The mode is the item which can be occurs more frequently

(ii) Discrete series: The mode is the value of the observation with greatest frequency

(iii) Continuous series:

$$Mode = l + \frac{(f - f_1)C.I.}{2f - f_1 - f_2}$$

Where $\frac{N}{2}$ = the median class where $N = \sum_{i=1}^{n} f_i$

l = the lower limit of the median class

f = the frequency of the median class

f_1 = the frequency of the class preceding the median class

f_2 = the frequency of the class succeeding the median class

$C.I.$ = the class interval of the median class

Relation between mean, median and mode: $Mode = 3\ Median - 2\ Mean$

Measures of Dispersion:

(i) Range: The range is the difference of the largest and smallest observations

Coefficient of range = $\frac{Range}{Maximum + Minimum}$

(ii) Mean deviation (M.D.): In the case of individual series, Let $x_1, x_2, x_3, \ldots \ldots, x_n$ be the n values of the variable x. Then the mean deviation about a point M is given by

$$M.D. = \frac{\sum |x_i - M|}{n}$$

In the case of individual series, Let $x_1, x_2, x_3, \ldots \ldots, x_n$ be the n values of the variable x and $f_1, f_2, f_3, \ldots \ldots, f_n$ be the corresponding frequencies. Then the mean deviation about a point M is given by (where M is the mean or median or mode)

$$M.D. = \frac{\sum f_i |x_i - M|}{N} \quad where\ N = \sum_{i=1}^{n} f_i$$

Coefficient of mean deviation = $\frac{Mean\ deviation}{M}$ where M is the mean or median or mode

(iii) Quartile deviation (Q.D.): If Q_3, Q_1 are the third quartile and first quartile then

$$Q.D. = \frac{Q_3 - Q_1}{2}$$

Individual series: Arrange the items in ascending or descending order. Then

$$Q_1 = Size\ of\ \left(\frac{N+1}{4}\right)^{th} item\ and\ Q_3 = Size\ of\ \left(\frac{3(N+1)}{4}\right)^{th} item$$

Discrete series: Arrange the items in ascending or descending order and find the cumulative frequencies of the items. Then

Q_1 = The cumulative frequency just more than $\frac{N}{4}$ and

Q_3 = The cumulative frequency just more than $\frac{3N}{4}$ where $N = \sum_{i=1}^{n} f_i$

Continuous series:

$$Q_i = l + \frac{\left(i\frac{N}{4} - F\right)C.I.}{f}\ for\ i = 1,2,3$$

Where l = the lower limit of the class whose cumulative frequency just greater than

$i\frac{N}{4}$

f = the frequency of the class

F = the cumulative frequency of the class preceding the class

$C.I.$ = the class interval of the class

Coefficient of quartile deviation = $\frac{Q_3 - Q_1}{Q_3 + Q_1}$

(iv) Variance (σ^2):

In case of individual series,

$$\sigma^2 = \frac{1}{n}\sum_{i=1}^{n}(x_i - \bar{x})^2 = \frac{1}{n}\sum_{i=1}^{n}x_i^2 - \left(\frac{1}{n}\sum_{i=1}^{n}x_i\right)^2$$

In case of discrete and continuous series,

$$\sigma^2 = \frac{1}{N}\sum_{i=1}^{n}f_i(x_i - \bar{x})^2 = \frac{1}{N}\sum_{i=1}^{n}f_i x_i^2 - \left(\frac{1}{N}\sum_{i=1}^{n}f_i x_i\right)^2$$

$$where\ N = \sum_{i=1}^{n} f_i$$

(v) Standard deviation (S.D.): The positive square root of variance is called as the standard deviation.

Coefficient of standard deviation $= \dfrac{\sigma}{\bar{x}}$

Coefficient of variation $= \dfrac{\sigma}{\bar{x}} \times 100$

Relation between Q.D., M.D., S.D.: $Q.D. < M.D. < S.D.$

Combined variance: If $\bar{x_1}, \bar{x_2}$ are the means and σ_1, σ_2 are the standard variations of two samples of sizes n_1 and n_2 then the combined variance is

$$\sigma^2 = \dfrac{n_1(\sigma_1{}^2 + d_1{}^2) + n_2(\sigma_2{}^2 + d_2{}^2)}{n_1 + n_2}$$

Where $d_1 = \bar{x_1} - \bar{x}$ and $d_2 = \bar{x_2} - \bar{x}$ and $\bar{x}$ is the combined mean

53. PROBABILITY

Random Experiment: The experiment is called as a random experiment when the total outcomes of the experiment well known in advance and the actual outcome doesn't know in advance.

Eg: Tossing a coin, Throwing a die

Sample space (S): The set of total outcomes of a random experiment is called as a sample space of the experiment. It is denoted by S.

Eg: The sample space when throwing a fair die is $S = \{1,2,3,4,5,6\}$

Elementary event or Event: Any subset of sample space is called as the event.

Eg: Let E be the evnt of getting an even number when throwing a fair die

i.e. $E = \{2,4,6\}$

Complementary event: The event doesn't occur is called as the complementary event of the event which occur.

Let A be the event which occur then the complementary event of A is denoted by A^c or $\bar{A}$ and is defined as $A^c = S - A$

Mutually exclusive events: Two events of a random experiment are said to be mutually exclusive events when no element common for the two events

Let A and B be two events of a random experiment then they are said to be mutually exclusive events when $A \cap B = \emptyset$

Eg: The events A and B of getting an even number and an odd number respectively when throwing a die are mutually exclusive events

$A = \{2,4,6\}$; $B = \{1,3,5\}$; $A \cap B = \emptyset$

Equally likely events: Two or more events of a random experiment are said to be equally likely events when they have same number of elements.

Eg: The events A and B of getting an even number and an odd number respectively when throwing a die are equally likely events

$A = \{2,4,6\}$; $B = \{1,3,5\}$; $n(A) = n(B) = 3$

Exhaustive events: Two or more events of a random experiment are said to be exhaustive events when their union is a sample space.

Eg: The events A and B of getting an even number and an odd number respectively when throwing a die are exhaustive events

$A = \{2,4,6\}$; $B = \{1,3,5\}$; $A \cup B = S$

Favourable cases: The number of outcomes to favourable to the event that calculate the probability is called as the favourable cases of the event

Classical definition of probability: The probability of getting the event A is denoted by $P(A)$ and is defined as $P(A) = \dfrac{n(A)}{n(S)}$

Limitations of probability: Let A be a event of a random experiment. Then $0 \leq P(A) \leq 1$

Certain event and impossible events: Let A be a event of a random experiment. Then

(i) A is called certain event when $P(A) = 1$

(ii) A is called impossible event when $P(A) = 0$

Odds in favour and odds in against: Let A be the event of a random experiment. Then

The odds in favour of A $= \dfrac{P(A)}{P(\bar{A})}$

The odds in against of A $= \dfrac{P(\bar{A})}{P(A)}$

Classification of pack of cards:

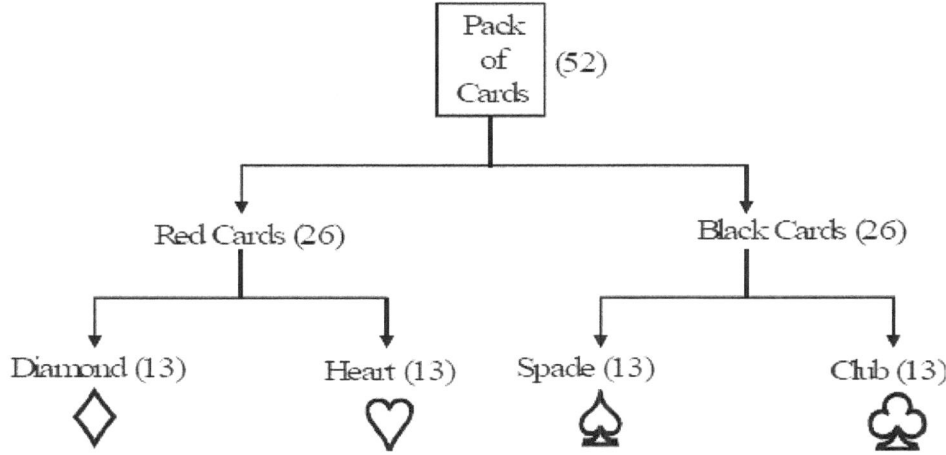

(i) Diamonds, Hearts, Spades, Clubs are called as suits

(ii) Each suit has contains 13 cards namely 2,3,4,....,10,J,Q,K,A

(iii) J,Q,K are called the face cards or court cards

(iv) There are 12 face cards in the pack

(v) J,Q,K,A are called the honoured cards

(vi) There are 16 honoured cards in the pack

(vii) J is called the knave card

(viii) There are 4 knave cards in the pack

(ix) There are the number of cards with same same value is 4 in the pack

(x) There are 36 number cards in the pack

Set Notations: Let A and B be two events of a random experiment. Then

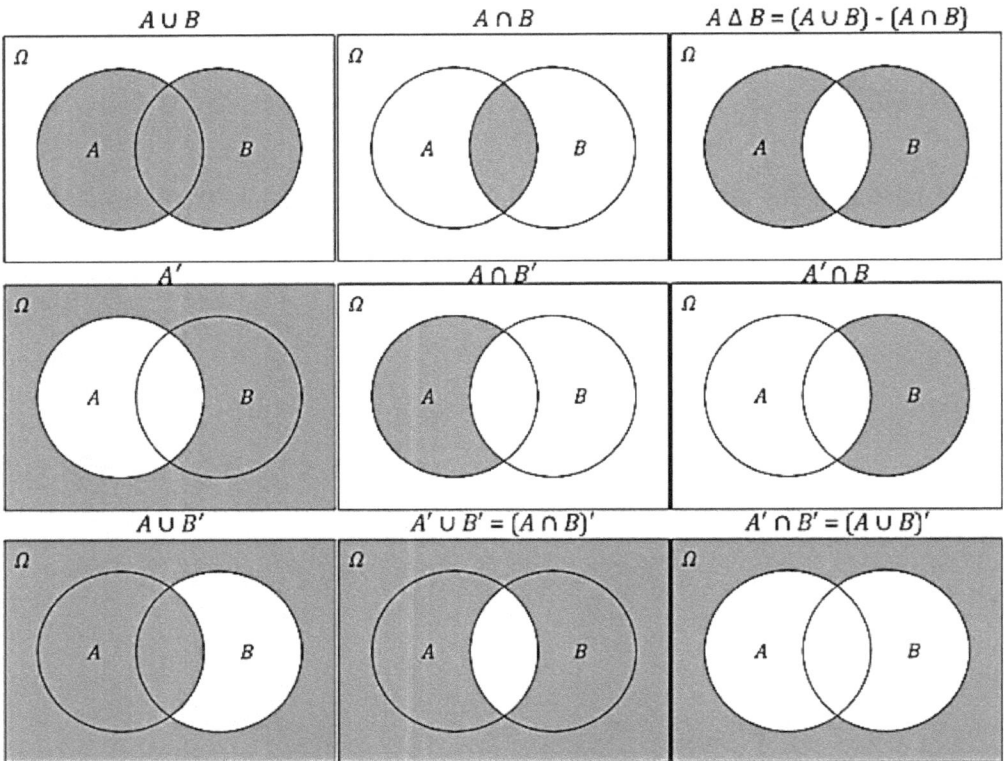

Addition theorem: Let A and B be two events of a random experiment then

$P(A \cup B) = P(A) + P(B) - P(A \cap B)$

Properties:

(i) If $A \subseteq B$ then $P(A) \leq P(B)$

(ii) $P(A - B) = P(A) - P(A \cap B)$

(iii) $P(A \cap B) \leq P(A)$ or $P(B) \leq P(A \cup B) \leq P(A) + P(B)$

(iv) $P(A \cup B \cup C) = P(A) + P(B) + P(C) - P(A \cap B) - P(B \cap C) - P(C \cap A) + P(A \cap B \cap C)$

(v) $P(A^c \cap B^c) = 1 - P(A \cup B)$

(vi) $P(A^c \cup B^c) = 1 - P(A \cap B)$

(vii) If A, B, C are mutually exclusive and exhaustive events of a random experiment then $P(A \cup B \cup C) = P(A) + P(B) + P(C) = P(S) = 1$

Conditional event: The occurrence of an event after the occurrence of the other event is called as the conditional event.

Conditional Probability: Let A and B be the two events of a random experiment. Then the probability of occurrence of A after the occurrence of B with $P(B) \neq 0$ is called conditional probability of A given B and is denoted as $P(A/B) = \dfrac{P(A \cap B)}{P(B)}$

Multiplication theorem: Let A and B be two events of a random experiment with $P(A) \neq 0$ and $P(B) \neq 0$ then $P(A \cap B) = P(A/B)P(B) = P(B/A)P(A)$

Independent events: Let A and B be the two events of a random experiment. Then the events are called as independent events when $P(A \cap B) = P(A).P(B)$

Properties: Let A and B be the two independent events of a random experiment. Then

(i) A^c and B^c are independent events i.e. $P(A^c \cap B^c) = P(A^c).P(B^c)$

(ii) A^c and B are independent events i.e. $P(A^c \cap B) = P(A^c).P(B)$

(iii) A and B^c are independent events i.e. $P(A \cap B^c) = P(A).P(B^c)$

Baye's Theorem: Let $E_1, E_2, E_3, \ldots \ldots, E_n$ be n mutually exclusive and exhaustive events of a random experiment with $P(E_i) > 0$ for $i = 1,2,3,\ldots.,n$. Let A be any event other than these n events. Then for any event E_k of the events of the random experiment

$$P(E_k/A) = \frac{P(E_k).P(A/E_k)}{\sum_{i=1}^{n} P(E_i).P(A/E_i)}$$

54. RANDOM VARIABLES AND PROBABILITY DISTRUBUTIONS

Random Variable: Let S be a sample space of a random experiment. Then the real valued function $X: S \to R$ is called as a random variable.

Discrete Random Variable: If a discrete variable X take the values $x_1, x_2, x_3, \ldots \ldots, x_n$ with respective probabilities $P(x_1), P(x_2), P(x_3), \ldots \ldots, P(x_n)$ such that $P(x_i) \geq 0 \; \forall i$ and $\sum_{i=1}^{n} P(x_i = 1)$ then X is said to be discrete random variable.

Probability distribution: The set of ordered pairs $\{x_i, P(x_i)\}$ is called as the probability distribution of a discrete random variable X.

Mean (μ or $\bar{x}$): The expected value or average or mean or first moment about origin is

$$E(x) = \bar{x} = \mu = \mu_1{}^1(0) = \sum_{i=1}^{n} x_i P(x_i)$$

Second moment about origin:

$$E(x^2) = \mu_2{}^1(0) = \sum_{i=1}^{n} x_i{}^2 P(x_i)$$

Variance:

$$\sigma^2 = \mu_2 = \sum_{i=1}^{n} x_i{}^2 P(x_i) - \mu^2$$

Cumulative distribution function: If X is a discrete random variable. Then the sum of the probabilities of the values less than or equal to a particular value x_k is called the cumulative distribution function of the value x_k and it is denoted by $P(X \leq x_k) = \sum_{i=1}^{k} P(X = x_i)$

Note: (i) If the mean of the random variable X is $\bar{x}$ then the mean of the random variable $aX \pm b$ is $a\bar{x} \pm b$ where a and b are real numbers.

(ii) If the variance of the random variable X is σ^2 then the variance of the random variable $aX \pm b$ is $a^2 \sigma^2$ where a and b are real numbers.

BINOMIAL DISTRIBUTION

Bernoulli Trials: The random trials which result either in the success or failure of an event with constant probability of success p and that of failure $q = 1 - p$ are called Bernoulli trials.

Binomial distribution: The probability of r successes in n independent Bernoulli trials is given by $P(X = r) = {}^nC_r p^r q^{n-r}$; $p \geq 0, q \geq 0$ and $p + q = 1$ is called the binomial distribution. It is also denoted by $B(k; n, p)$. Here n, p are called the parameter

Symmetrical binomial distribution: If $p = q = \frac{1}{2}$ then the binomial distribution is called as the symmetrical binomial distribution.

Mean: $\bar{x} = np$

Variance: $\sigma^2 = npq$

Standard deviation: $\sigma = \sqrt{npq}$

Mode: The mode is the value of variable with maximum probability. It depends on the value of $np + p$

$$Mode = \begin{cases} k \text{ if } [np + p] = k \text{ and } np + p \text{ is not an integer} \\ k \text{ and } k - 2 \text{ if } [np + p] = k \text{ and } np + p \text{ is an integer} \end{cases}$$

Note: (i) If μ and σ^2 are the mean and variance of a binomial distribution then $0 \leq \sigma \leq \sqrt{\mu}$

(ii) In binomial distribution, $P(X = r)$ is maximum when $r = [np]$

(ii) If we conduct n independent Bernoulli trials repeated N times the expected frequency of r successes is $f(X = r) = N \cdot {}^nC_r p^r q^{n-r}$

POISSON DISTRIBUTION

Poisson distribution: The probability of r successes in n independent Bernoulli trials is given by $P(X = r) = \frac{e^{-\lambda} \lambda^r}{r!}$ where $\lambda > 0$, parameter is called as a poisson distribution.

Mean=Variance=parameter= λ

Standard deviation=$\sqrt{\lambda}$

If we conduct n independent Bernoulli trials repeated N times the expected frequency of r successes is $f(X = r) = N \frac{e^{-\lambda} \lambda^r}{r!}$

KASTURI VIJAYAM

📞 00-91 95150 54998
KASTURIVIJAYAM@GMAIL.COM

SUPPORTS

- PUBLISH YOUR BOOK AS YOUR OWN PUBLISHER.

- PAPERBACK & E-BOOK SELF-PUBLISHING

- SUPPORT PRINT ON-DEMAND.

- YOUR PRINTED BOOKS AVAILABLE AROUND THE WORLD.

- EASY TO MANAGE YOUR BOOK'S LOGISTICS AND TRACK YOUR REPORTING.

www.ingramcontent.com/pod-product-compliance
Lightning Source LLC
LaVergne TN
LVHW070024080526
838202LV00066B/6818